T0155643

Instruction Level Parallelism

Alex Aiken • Utpal Banerjee • Arun Kejariwal • Alexandru Nicolau

Instruction Level Parallelism

Alex Aiken
Department of Computer Science
Stanford University
Stanford, CA, USA

Utpal Banerjee
Department of Computer Science
University of California at Irvine
Irvine, CA, USA

Arun Kejariwal
MZ Inc.

Alexandru Nicolau
Center for Embedded Computer Systems
University of California at Irvine
Irvine, CA, USA

ISBN 978-1-4939-7959-2 ISBN 978-1-4899-7797-7 (eBook)
DOI 10.1007/978-1-4899-7797-7

Printed on acid-free paper

This Springer imprint is published by Springer Nature
The registered company is Springer-Verlag London Ltd.

This book is dedicated to our families.

Contents

List of Figures

List of Tables

Preface

This book has been distilled over more than a decade. It started as part of a more ambitious project that Utpal Banerjee and Alex Nicolau discussed over a rather ample dinner around the turn of the century. The book, as originally planned, would have consisted of several volumes culminating in a unification of all parallelizing compiler technology in an all-encompassing framework/formalization. Over time, the project gained a sense of proportion, and was reduced to the more realistic goal of a thorough description of Instruction Level Parallelism and a more complete, unifying formalization of the techniques within. After an initial period of re-education about the subject, gathering and agreement on the most representative material to include followed by (very) long, detailed discussions to ascertain that both authors had a common understanding of the material, we finally started writing, some 4–5 years after the original discussion of the project. This writing phase took an additional two years, and resulted in four chapters being written in varying degrees of detail (Basic Block techniques, Definitions, Architectures in close-to, but not quite ready for publication form, and Trace Scheduling re-formalized by Utpal with some interesting new results in the bargain). The rest of the chapters ranged from detailed outlines with some sections in rough form, to little more than titles and subsection titles.

We then decided to add an extra author, who with fresh energy and optimism, would help write the remaining chapters and have the stamina to carefully read, fix, expand and homogenize the whole text. In this way, Arun Kejariwal joined the team, and after six months or so of reading, he quickly (one year) produced important additions to the work, including (but not limited to) an encyclopedic collection of references, Further Study sections for each chapter and the preliminary work on the Loop Scheduling chapters. At that point, while the manuscript had grown to about 180 pages, it was not in what any of the authors would have considered close to publication-ready. We invited Alex Aiken to join the project owing to his deep understanding and contributions to research in Instruction Level Parallelism. Much to our surprise, Alex graciously agreed to join and not content to just help complete the Loop Scheduling chapters, which we saw as the minimal required effort to make the manuscript publishable, he volunteered to perform a very thorough pass on the whole manuscript and rewrite/revise/unify/amplify/simplify as needed. The resulting book is a much better end product for Alex's efforts. Indeed, the resulting book, after a total of approximately 15 years, is a team effort that we can all be proud of, and, realistically, is also much better than any of us could have produced individually. We can only hope that the reader will agree and find the book useful.

A. AIKEN, U. BANERJEE, A. KEJARIWAL , A. NICOLAU

Foreword

For more than fifty years, the most powerful processors have taken advantage of the parallelism between instructions. Most likely, this will continue to be the case since the constraints imposed by power have made exploitation of parallelism in all its forms, including at the instruction level, the most important strategy for continued advances in performance. Compiler techniques are important components of this strategy. This is particularly true of the techniques for instruction level parallelism which, in contrast to coarse grain parallelism, cannot be easily controlled by the programmer using high level language directives. Indeed, for VLIW processors, compiler techniques are necessary to detect and exploit instruction level parallelism and although the more popular superscalar processors incorporate powerful runtime mechanisms for this purpose, compiler strategies have the potential to enhance their effectiveness.

Because of the importance of compilers for instruction level parallelism, the early studies of five decades ago were rapidly accepted as an appealing area of study which has counted with the participation of numerous researchers and has produced an extensive literature. Surprisingly, there was, until now, no single document covering in detail and rigorously the most important compiler strategies for instruction level parallelism. Today's compiler textbooks contain a limited coverage and the treatment is somewhat superficial. Those wanting to develop expertise in the subject have been forced to navigate the literature without much guidance.

This monograph is the first attempt to remedy this situation. Its authors, Aiken, Banerjee, Kejariwal, and Nicolau, are particularly qualified to undertake this task. They are influential researchers who created some of the most influential ideas in instruction level parallelization. Building on this expertise, they wrote this valuable contribution to the compiler literature. For the first time, there is in a single document a description of the core strategies for instruction level parallelization complemented with an extensive guide to the literature on the subject. Because of the elegance of the strategies and theory presented, I expect this book will serve not only to expand the knowledge of its readers but also as a source of inspiration for researchers.

D. PADUA
ACM/IEEE Fellow

Acknowledgments

Partly due to the unusually long gestation of this book, preliminary versions for parts of three of the book chapters have appeared in print previously. The Trace Scheduling re-formalization was published as an ACM TOPLAS paper, and preliminary descriptions of Basic Block Scheduling and Software Pipelining appeared in the Encyclopedia of Parallel Processing.

We owe an enormous debt of gratitude to Michael Wolfe, who did an amazingly detailed job of proofreading the manuscript and pinpointing many problems, omissions and errors in the text. The resulting book is a much better work due to his selfless and heroic efforts, though of course any remaining deficiencies are the exclusive fault of the authors. Last but not least, we would like to thank our Springer Senior Editor, Melissa Fearon, for her forbearance and encouragement through the many years the completion of this volume has required.

1

INTRODUCTION

This introductory chapter discusses the role of instruction level parallelism (ILP) in optimizing compilers and in machine architectures that automatically reorder or parallelize programs. A brief overview of ILP is given along with an outline of the topics covered in this book.

1.1 Scope of the Book

Compilers have been central in making the task of computer programming more productive. By allowing the human programmers to deal with the machine at a higher level of abstraction, they have enabled the design of much more complex and longer programs than would have been otherwise feasible, as well as allowed portability and readability—sharing—of programs. Thus, compilers are arguably one of the most important factors in the widespread use of computers and their usefulness.

At first, however, it was not at all clear that compilers would be practical, as there was concern that too much performance would be sacrificed compared with hand-written code. In fact, the FORTRAN I compiler, the first compiler to become popular and establish compiler technology's potential, succeeded largely because it focused on providing performance close to what expert programmers could

A. Aiken et al., *Instruction Level Parallelism*,
DOI 10.1007/978-1-4899-7797-7_1

achieve on their own. Within a very short time after the debut of FORTRAN I, most programs were written in high-level languages (and in FORTRAN in particular) that were compiled to machine code.

The very popularity of compilers, however, started a new dynamic. Machines no longer had to be designed primarily to be programmed by hand and as a result they became increasingly sophisticated and complex, a trend that continues to this day. As machines evolved, a gap opened up in the relative performance of hand-coded and compiled code—it turned out that talented experts could find ways to exploit the machines that compilers could not. As a rule of thumb, a factor of 2 or 3 of lost performance came to be considered an acceptable tradeoff for the improved programmer productivity of writing in high level languages, particularly when the code in question was not performance critical.

However, for certain applications, speed is of the essence. These applications—coupled with advances in technology and computer architecture—have led to the development of novel architectures such as Pipelined, Vector, Superscalar/VLIW, and Parallel Multiprocessors, each of which uses a form of parallelism to improve performance. While for a simple sequential machine the loss of speed due to poor code generation was relatively modest, the penalty for poor code generation for a parallel machine can be proportional to the amount of parallelism available in the machine, even if the machine is relatively simple (e.g., a machine that captures temporal parallelism, such as a straightforward pipelined RISC processor). To make matters worse, the task of generating good parallel code by hand for most of these machines is much more tedious and error prone than the generation of sequential assembly code. Thus, the need for good optimizing compilers is even more critical, while the problems that need to be solved in designing such compilers are harder.

The compiler, however, is only one factor in the overall scheme to obtain good performance. One must start with a good algorithm. The programming language should allow the natural expression of the algorithm, and the architecture should support an efficient implementation. Finally, the operating system must be capable enough to handle a satisfactory execution. While much research has gone into optimizing each one of these factors individually, there is still no clear consensus on the ideal language, compiler, architecture, and OS combination (we leave algorithmic development out of the scope of this book).

While there clearly is a very tight relationship between the language, compiler, and architecture, among these the language is perhaps the one most given to subjective choices. In this book, we deal only with compilers, and discuss language/machine features only in as much as they affect the compiler technology being discussed. Furthermore, the book primarily deals with compiling for instruction-level parallelism (ILP). To make the discussions and algorithms presented here generally useful, we do not describe them in terms of a particular commercial product, but rather in terms of a well-defined abstract model that represents the class of machines to which the techniques apply. Thus, the book is organized around major compilation techniques, not around architectural groups of machines now in existence, though we try to make the relationship between the techniques and the machines to which they apply quite clear. We also comment, whenever appropriate, on the implications of different languages on the compiler. However, for most of the book, we assume that the high level language used is the predominant one typically used for that class of machines.

With this approach, we hope to provide a useful reference for the design of a state-of-the-art parallelizing compiler for ILP machines (machines that can exploit instruction-level parallelism). We also give a broader view of the tradeoffs between the various compiler and architectural issues and approaches that should carry over to future generations of machines as well as being immediately applicable to current designs.

1.2 Instruction-Level Parallelism

In recent years we have seen a proliferation of machines seeking to exploit *instruction-level parallelism* (ILP), that is, parallelism at the machine instruction level. Instruction-level parallelism is achieved by two techniques, which may be used separately or in combination:

- Multiple instructions may be issued at the same time, requiring multiple functional units (such as adders and multipliers) that can each execute an instruction simultaneously. The two primary styles of multiple issue machines are *superscalars*, where the decisions about what instructions to issue are made in hardware at runtime, and *VLIWs*, where the decisions are made in advance by a compiler.

- Instructions can be *pipelined*, meaning that individual instructions are subdivided into a sequence of stages each of which is handled separately in a fixed order. For example, instructions might be broken up into a *decoding* stage (determining what kind of instruction is being executed), a stage to load data needed by the instruction (if the instruction reads from memory), a stage to perform the operation itself, and a stage to write results back (if memory stores are required). As soon as an instruction is done with the first stage and has moved on to the second stage, another instruction can begin the first stage. In this was, pipelining allows partial overlap of instructions and in general an n-stage pipeline allows up to n instructions (each using a different stage of the pipeline) to be executing simultaneously. Increasing the number of stages allows finer-grain overlap and a faster clock cycle.

The popularity of instruction-level parallelism been fueled primarily by technological advances that have made architectures incorporating multiple function units and high bandwidth instruction issue feasible, even in single-chip microprocessors. Essentially every processor produced today utilizes instruction pipelining, multiple issue or, most commonly, both.

Instruction-level parallelism is, therefore, the dominant architectural approach to high performance CPU design, and will continue to be so for the foreseeable future. Furthermore, as these architectures grow larger, the need to match this technological advance with the corresponding increase in the parallelism *actually exploited from applications* becomes critical. Thus, efforts to aggressively expose such parallelism in application programs are increasingly becoming important, if the overall system performance is to keep up with the parallelism promised by the technology advances. In this context, compiler extraction and exploitation of instruction-level parallelism is a central task.

It should be noted that compilation techniques for exploiting ILP are useful regardless of the sophistication of the hardware. At compile time a relatively global search for parallelism, including sophisticated analysis of the code and complex and time consuming transformations, can be undertaken that would be impractical to do at runtime, either in hardware or software. However, in the past, the need for aggressive compilation techniques has sometimes been discounted, to the point that compilers for aggressively superscalar architectures have often produced code that is substantially worse that might be expected. These suboptimal programs do not always result in significant degradation of performance, partly due to the relatively limited parallelism available in the hardware and partly due to the sophisticated runtime scheduling of the superscalar aspects of the architecture. However, current generations of ILP machines find it harder to keep increasing performance while relying mainly on aggressive superscalar features (e.g., renaming multilevel prediction) without degradation of the clock cycle. These limits coupled with the wider issue bandwidth and increased numbers of functional units available make closer coordination with aggressive compiler technology even more critical for overall performance improvement in future machines.

Instruction level parallelism was first studied in the context of microcode compaction, which is not surprising because horizontal microcode is a primitive form of parallelism and it was the first technology developed where instruction level parallelism was truly critical. Microcode has several characteristics that are markedly dissimilar from those of machine language code. Most importantly, microcode conditionals tend to be highly biased towards one or the other of the branches. Nevertheless, microcode, like machine code, is rich in branches, and thus can benefit from the instruction scheduling techniques that handle branches well.

1.3 Outline of Topics

We begin in Chapter 2 with a historical overview of support for ILP in machine architectures from the first electronic computers up through the present day, illustrating the increasing exploitation of ILP over time as well as illustrating the tradeoffs that are made to design systems that are as efficient as possible given the shifting constraints of technology over time.

The main material of the book, compilation techniques for ILP, is presented beginning with the simplest techniques and progressing to the most sophisticated, which also happens to correspond very closely to the historical order in which these techniques were developed. Chapter 3 begins with *basic block techniques* for extracting ILP from straight-line sequences of code. This problem is both fundamental and of practical importance in its own right. The ideas presented in this chapter are also reused, both explicitly and implicitly, in techniques developed subsequently for more complex kinds of programs.

Trace scheduling was the first viable method for instruction scheduling beyond basic blocks and had a major influence on all subsequent work in compiling for ILP. Chapter 4 presents trace scheduling and a number of extensions in detail, including new, simpler arguments proving the correctness of trace scheduling's fairly involved approach to exploiting ILP.

Efforts to simplify and generalize trace scheduling led directly to the development of *percolation scheduling,* which is presented in Chapter 5. Percolation scheduling is a program transformation system consisting of a few primitive program transformations. Because of its generality and the fact that any sequence of the primitive transformations is guaranteed to be correct, percolation scheduling has been used in numerous subsequent publications as the framework for explaining a higher-level algorithm that uses the primitive percolation scheduling transformations as building blocks.

Early techniques for exploiting ILP in loops amounted to unrolling the loop some number of times and then applying an algorithm such as trace scheduling to the loop body. Besides incurring code blowup, such approaches fail to exploit a considerable amount of the parallelism available in loops. The invention of the first *software pipelining* technique, called *modulo scheduling,* that could potentially exploit all of the available ILP across all iterations of a loop, was a major advance. Modulo scheduling is presented in Chapter 6.

Another class of software pipelining techniques, based on *kernel recognition,* overcome a number of limitations of modulo scheduling and have stronger theoretical properties, including in some cases (and under some assumptions) the ability to generate provably optimal code. Kernel recognition techniques are covered in Chapter 7.

FURTHER READING

For basic concepts and classical compiler optimizations the reader is advised to refer to any one of books by Aho et al. [ALSU06], Barrett [BBGC86], Muchnick [Muc00], Allen and Kennedy [AK01] or by Cooper and Torczon [CT03]. The extent of ILP available for superscalar and superpipelined machines was discussed by Jouppi and Wall [JW89]; the effect of non-uniform distribution of ILP on performance is discussed in [Jou89]. Limits on multiple instruction issue for exploiting ILP is discussed by Smith et al. in [SJH89]; a very good discussion of the limits of ILP is available in [Wal91]. An excellent reference for a historical overview of ILP is the survey paper by Rau [RF93]. A review of ILP for instruction-level parallel processors is available in [FFY01].

2

OVERVIEW OF ILP ARCHITECTURES

In this chapter we trace the history of computer architecture, focusing on the evolution of techniques for instruction-level parallelism. After briefly summarizing the early years of machine design, we focus on the development of out-of-order, pipelined, and multiple-issue processors. These are further divided into processors that do instruction scheduling entirely in hardware (e.g., superscalar machines) and those that expose the instruction scheduling to the compiler, particularly VLIW machines such as the Multiflow Trace, Cydra 5, and Itanium.

2.1 Historical Perspective

The primitive form of ILP, called *horizontal microcode* in current terminology, first appeared in Turing's 1946 design of Pilot ACE [CTW86], further discussed by Wilkes in [Wil51, WS53]. The initial footsteps towards parallel processing can be traced back to the Livermore Atomic Research Computer (LARC) [ECTS59] and the IBM 7030 (a.k.a. STRETCH) [Blo59] that overlapped memory operations with processor operations. The 7030 embodied several new architectural concepts, including a fully pipelined system with highly interleaved memory and speculation on branch as a bit set in the branch instruction. Other important commercial machines of the 1950s include the

A. Aiken et al., *Instruction Level Parallelism*,
DOI 10.1007/978-1-4899-7797-7_2

IBM 704 and its successors, the 709 and 7094. The latter introduced I/O processors for better throughput between I/O devices and main memory.

In the early 1950s Engineering Research Associates (ERA) developed the transistorized ERA 1103 computer. In 1952 ERA was sold to Remington Rand, an early American computer manufacturer, best known as the original maker of the UNIVAC; Rand soon merged with Sperry Corporation. In 1957 some of the ERA founders, including William Norris and Seymour Cray, left Sperry to form CDC. In 1959 CDC released a 48-bit 1103 design as the CDC 1604. A 12-bit cut down version was also released as the CDC 160 in 1960, arguably the first minicomputer. In the early 1960s new versions of the basic 1604 architecture were re-built into the CDC 3000 series. In 1964 the CDC 6600 was released, which outperformed everything on the market at that time by roughly a factor of ten. The 6600 had a simple CPU, but used a series of external I/O processors to offload many common tasks, enabling the CPU to devote all of its time and circuitry to processing data while the other controllers dealt with tasks like punching cards and running disks. In 1969 CDC released the 7600 machine, which ran about four to five times the speed of the 6600; much of the performance improvement came from extensive use of pipelining.

In 1965 DEC developed the IMS 6100, a single chip design of the PDP-8 (Programmed Data Processors) minicomputer, a successor of the PDP-5 (1963). The 6100 was a 12-bit processor, which had exactly three registers: the PC, AC (an accumulator), and MQ. All 2 operand instructions read AC and MQ, and wrote back to AC. It had a 12-bit address bus, limiting RAM to only 4K and had no stack. Memory references were 7-bit (128 word) offset either from address 0 or the PC. The PDP-8 was succeeded by the PDP-11, designed in part by Gordon Bell. The PDP-11 was succeeded by the 32-bit VAX (Virtual Address Extension) architecture.

The 1970s saw the emergence of pipelined computers [Dav71, Sha72, DSTP75, Pat78]. Most notably, Davidson's group developed the theory of and algorithms for the design of pipelined machines [Dav74, TD74]. Patel and Davidson proposed the insertion of delays [PD76] for improving pipeline throughput; a discussion on the microprogramming of and algorithm development for pipelined processors by Kogge is available in [Kog77b] and [Kog77a] respectively. In [Lar73] Larson discussed the optimal number of pipeline stages.

Superscalar and superpipelining techniques are discussed in [JW89]. For further insight into the architecture of the early pipelined computers, the reader is advised to refer to [CK75, RL77, Kog81]. Scheduling algorithms for pipelined algorithms are reported in [BJS80, Rym82, GM86, SP89]. A survey of the instruction scheduling techniques for pipelined processors is available in [Smi89]. Techniques for code generation and code optimization for pipelined architectures are discussed in [HG82] and [Gro83] respectively.

The 1970s also saw the development of multiple processors and vector hardware for parallelism. The SOLOMON computer [GM63], developed by Westinghouse Corporation, and the ILLIAC IV [BBK+68, Dav69], jointly developed by Burroughs, the Department of Defense and the University of Illinois, were representative of the first parallel computers. The Texas Instrument Advanced Scientific Computer (TI-ASC) [Wat72] and the STAR-100 of CDC [HT72] were pipelined vector processors that demonstrated the viability of that design and set the standards for subsequent vector processors. At this time, Cray, Inc. introduced high performance vector supercomputers that had strong processors, high speed memory systems and powerful I/O systems. In 1976, the CRAY-1 [Rus78] became the supercomputer industry standard. Since vector parallelism is mostly distinct from ILP, we omit further discussion of vector processors in the rest of the book. The reader is advised to refer to the survey by [Don86] and the book by Schneck [Sch87] for a detailed description of vector architectures.

The idea of data-flow computing, and in particular data-flow graphs, originated from the formal studies in the mid-1960s [KM66, Ada68, RB69]. Data-flow computer architectures are based on the principle of activating an instruction contingent on the availability of the inputs the instruction needs. A distinctive feature of data-flow computers is that they do not have the program counter characteristic of von Neumann computers; also, they do not have addressable (i.e., random-access) memory. Dennis and Misunas first introduced an architecture for a data-flow processor in the early 1970s [DFL74, Den74, DK82] as well as an early data-flow machine [DM75]; other notable efforts were the LAU project at Toulouse, France [Pla76] and the DDM1 machine [Dav78]. The two forms of data-flow architecture are the *tagged-token architecture* and *static architecture*. The former provides hardware support for function application (including recursion) and overlapped execution of successive loop iterations, whereas the latter requires compiler support for these programming features. The static model was best suited for regular numerical applications, whereas the tagged-token model was better suited to applications with less predictable runtime behavior, such as symbolic manipulation. Examples of the static data-flow architecture include [Cor79, THH80, HNI82, DG83]; examples of tagged-token model include [AK81] [WG82, GKW85, GDHH89, PC90]. Since data-flow architectures are not representative (in the true sense) of ILP computing, we omit further discussion of data-flow machines from this book. An overview of the evolution of the data-flow architectures is available in [Den91] and a survey of data-flow architectures is available in [Vee86].

In 1971 Intel developed the first single chip CPU, the Intel 4004 – a 4-bit processor meant for a calculator. It processed data in 4 bit units, but its instructions were 8 bits long. Program and data memory were separate, with 1K of data memory and a 12-bit PC for 4K program memory (in the form of a 4 level stack, used for CALL and RET instructions). There were also sixteen 4-bit (or eight 8-bit) general purpose registers. The 4004 had 46 instructions, using only 2,300 transistors in a 16-pin DIP and ran at a clock rate of 740kHz. The 4040 (1972) was an enhanced version of the 4004, adding 14 instructions, larger (8 level) stack, 8K program space, and interrupt abilities (including shadows of the first 8 registers). In 1973 Intel introduced the 8-bit 8008 processor.

In 1976 Texas Instruments developed one of the first true 16-bit microprocessors, the TMS 9900 (the distinction of being first probably goes to the National Semiconductor PACE or IMP-16P, or AMD 2901 bit slice processors in 16-bit configuration). It was designed as a single chip version of the TI 990 minicomputer series, much like the Intersil 6100 was a single chip PDP-8, and the Fairchild 9440 and Data General mN601 of Data General's Nova. The TMS 9900 had a 15-bit address space and two internal 16-bit registers. One unique feature was that all user registers were actually kept in memory, including stack pointers and the program counter. A single workspace register pointed to the 16 register set in RAM, so when a subroutine was entered or an interrupt was processed, only the single workspace register had to be changed, unlike some CPUs which required a dozen or more register saves before acknowledging a context switch.

Similar to the PDP-8 (and IMS 6100), the Signetics 2650 (1978) was based around a set of 8 bit registers with R0 used as an accumulator, and six other registers arranged in two sets (R1A-R3A and R1B-R3B); a status bit determined which register bank was active. The other registers were generally used for address calculations (ex. offsets) within the 15 bit address range. This design kept the instruction set simple—all loads/stores to registers went through R0. It also had a subroutine stack of eight 15 bit elements, with no provision for spilling over into memory. A number of other 16-bit processors were introduced at time, including the ZILOG Z8000 and Motorola 68000.

Advances in semiconductor technology in the 1980s had a profound impact on the computer industry, triggering the emergence of VLIW machines, which were a natural outgrowth of horizontal microcode and marking the onset of the era of modern superscalar machines. The development of VLIW and superscalar machines during the 1980s and the 1990s is discussed in Sections 2.4 and 2.5.

2.2 Superscalar and VLIW Machines

The goal of ILP is to speed up a single program written for a sequential processor by discovering and exploiting parallelism between individual machine instructions. Different types of ILP architectures accomplish this task in different ways. In this chapter, we define the two most important types of processors that utilize ILP and briefly trace their historical development from the sixties to the present.

Our starting point is a sequence of machine operations (e.g., memory loads and stores, integer additions, floating-point multiplications, branches, etc.), which is typically produced by a compiler for a high-level programming language such as C++ or Java. Usually there are certain *dependences* between these operations, based on which memory locations and registers are read and written by instructions in the sequence. In ILP processing, multiple machine operations may be executed in parallel; in order to preserve the original behavior of the program, the operations must be executed in such a way that their dependences are not violated. Thus, in an ILP system (compiler plus hardware) the following sequence of basic steps is involved in the execution of any program:

1. Find the dependences of the given sequence of operations;

2. Find sets of mutually independent operations (i.e., those without dependences among them);

3. Decide how to execute each set of independent operations: when should the operations execute, on which functional units, and using which registers.

Different classes of ILP architectures are characterized by the point at which the compiler hands over this sequence to the hardware.

In a *superscalar machine* [AC87], all three basic steps are handled in hardware. The processor considers instructions within a *window* of the program—a window is just a contiguous subsequence of instructions of a fixed length. Instructions that come before the window have already executed; instructions after the window have yet to execute or even be considered by the processor. The instructions inside the current window can be in a variety of states, including currently executing and waiting to execute. In each machine cycle, the processor considers those instructions inside the window that are not yet executing and identifies any instructions that can be issued in the current cycle. The processor must make sure that any operations that are issued are independent of all operations that precede it in the window (whether these operations are executing or not). The processor also tracks what machine resources (e.g., functional units) are being used by currently executing instructions; even if no dependence prevents an instruction from being issued, if a resource the instruction needs is not available the instruction must be delayed and reconsidered in future cycles. As instructions complete execution they leave the start of the window and the next instructions in the program are added to the end of the window. Some of the early superscalar processors include Apollo's DN10000 [Apo88], the ZS-1 [Smi89], the DN10000TX [BCF+91], the Metaflow architecture [PSS+91], the SuperSPARC™ [BK92], the PA-RISC processor [DWYF92] and the Motorola 88110 [DA92].

A *VLIW (Very Large Instruction Word)* machine is at the other extreme: all of the basic steps are handled in software. The compiler creates a sequence of *long instructions*, where an instruction consists of several machine operations that are to be executed simultaneously. Static parallelization enables exploitation of parallelism beyond the fixed window of operations used in superscalar machines, subject to resource constraints. The compiler then directs the hardware to execute the operations on specific functional units and control steps.

VLIWs have long and continuing history of development and compete successfully in the processor marketplace. Notably, as early as 1989 Intel introduced a processor incorporating VLIW issue of instructions [KM89b]. In the embedded systems context, TI has had a long-running series of DSP processors (C6xxx, TM3282 and precursors) that use VLIW technology [ST09], while Philips Trimedia processors have found use in multimedia [vdWVD+05]. The SHARC DSP, by Analog Devices, is another example of a VLIW architecture that is commercially successful [Ana] as is the ST Microelectronics ST200 family based on the Lx architecture [FBF+00]. Tensilica's Xtensa LX2 processor's FLIX (Flexible Length Instruction eXtensions) is also VLIW-like [Cad]. The Infineon Carmel DSP is another VLIW processor core intended for SoC [Inf]. The Intel and HP IA-64 EPIC [Gep02, RAB+12]) Itanium, combines concepts derived from VLIW architectures with superscalar features. Finally, ELBRUS computers, started as a Soviet-era series of VLIW supercomputers e.g., [Elbb]. The company (Elbrus) still exists, and has recently announced a new multicore processor for release in 2015, the Elbrus 8C , on which little information is available in English [Elba] but which continue to utilize VLIW concepts within each core.

In this chapter, we describe some of the superscalar and VLIW machines that have been built over the years.

2.3 Early ILP Architectures

VLIW ideas originated in parallel microcode in computing's earliest days and in the first supercomputers. In 1964, Control Data Corporation (CDC) unveiled the 6600, considered by many to be the first supercomputer. In 1969, CDC offered the 7600, which had pipelined functional units. The IBM 360/91 was released during the same period; it employed instruction lookahead, separate floating point and integer functional units and a pipelined instruction stream. The IBM 360-195 was comparable to the CDC 7600, deriving much of its performance from a cache memory. In the 1970s, many attached array processors and dedicated signal processors used VLIW-like wide instructions in ROM to compute fast Fourier transforms and other algorithms. We now present a brief overview of some of the early ILP architectures.

CDC 6600

The first 6600 computer was delivered by Control Data Corporation in October, 1964. The CDC 6600 was a large-scale, solid-state, general-purpose computing system that broke away from the architecture of the IBM 7090, which was in widespread use at that time, and established a standard of performance against which machines were compared for years. The basic logic circuit used was a Direct-Coupled Transistor Logic Circuit (DCTL). The machine had a distributed architecture (central scientific processor supported by ten peripheral machines) and was a reduced instruction set (RISC) machine many years before the term was invented [PS81]. The CDC 6600 is credited with introducing many features which were novel at the time. It departed from tradition by using multiple function units, an interleaved memory, I/O peripheral processors with their own instruction sets and internal memory, and facilities for multiprogramming such as relocation registers. In this machine the hardware decided which operation to issue in a given cycle; its model of execution was that of superscalar processors.

The 6600 was a register-oriented machine. It had three sets of eight registers: eight 18-bit A registers for addressing, eight 18-bit B registers for indexing, and eight 60-bit X registers for arithmetic/logical manipulation. This separation of arithmetic processing from memory access time was important since clock time was 100 ns and memory cycle time was 1,000 ns (1 ms). The routing of data among registers and functional units was managed by a data traffic control called the *scoreboard*. Instructions were issued in order, with delays caused only by the unavailability of a functional unit or the result register (not by unavailability of read operands).

A key description of the CPU is 'parallel by function.' The CPU had 10 functional units: boolean, branch, divide, fixed add, floating add, increment (2 units), multiply (2 units), and shift, which enabled the machine to perform a sequence of independent operations in consecutive cycles. The 6600 was a three address machine with two instruction formats. The 15-bit short form instruction had 6 bits for the

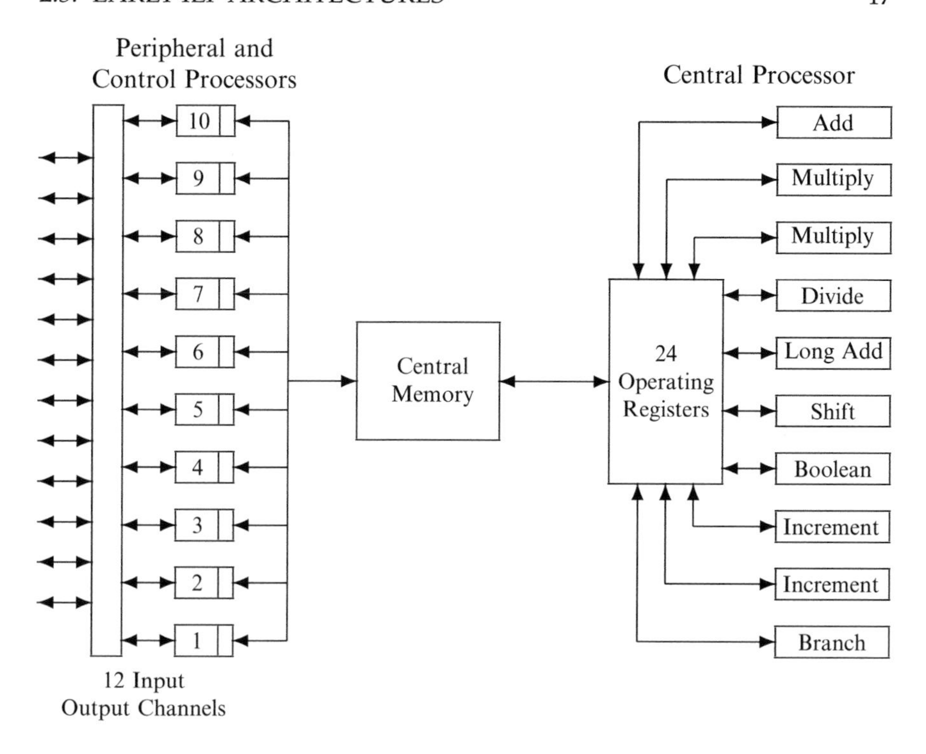

Figure 2.1 : CDC 6600 Block Diagram

opcode and 3 bits each for the result and the two operand fields. The result and operand fields specified the A, B, or X registers. A 30-bit long form instruction had 6 bits for opcode, 3 bits to specify a register to hold the result, 3 bits to specify a register for an operand, and 18 bits for an immediate operand. All computations were performed on quantities in registers; no arithmetic or logical instructions dealt with memory.

An instruction stack of eight 60-bit registers was used to hold a set of most recently executed instructions. An instruction fetched from the central memory first came to the stack's input register. Before it was accepted, the contents of the stack were shifted to push the oldest instruction out. This mechanism improved performance, since when control was transferred by a branch to an instruction already in the stack, that instruction did not need to be fetched from memory.

The central memory was divided into 32 banks of 4K 60-bit words. In the 17-bit address field, the low order 5 bits selected the bank and the high order 12 bits selected a word within a bank. Consecutive words resided in different banks. Since the banks operated independently, the memory responded to multiple requests at once. This interleaving provided greater bandwidth for randomly addressed operands and maximum bandwidth for sequences of consecutive operands. There were 10 peripheral processor units (PPU's) to carry out I/O and other external tasks. Each PPU was a small high-speed processor equipped with a local memory of 4K 12-bit words and with access to the central processor's main memory. Further discussion of the 6600 architecture is available in [Tho70].

IBM 360/91

The IBM 360/91 was a landmark in the development of instruction level parallelism. This fact was not recognized at the time, and in fact the algorithms developed by R. Tomasulo [Tom67] for hardware-based ILP exploitation were abandoned in later machines in the 70's and 80's. However, these algorithms, with those of the CDC 6600, form the core of all superscalar processors in use today. The machine adopted an assembly-line processing approach, using the techniques of storage interleaving, arithmetic concurrency, and buffering to provide high performance. Although the machine offered less ILP than the CDC 6600 (due to fewer functional units) it was far more ambitious than the CDC 6600 in its attempt to rearrange the instruction stream to keep the functional units busy—a key characteristic of superscalar machines.

The architecture of the IBM 360/91 is shown in Figure 2.2. The machine had no cache (the concept of caching was not yet recognized at that time) and relied on a 16-way interleaving of memory to increase memory bandwidth, allowing multiple memory accesses to proceed in parallel. Memory was thus divided into sixteen separate memory banks, each with its own decoding logic, and the overall addressing was designed to allow consecutive addresses to fall in separate banks. When an access request (load) was gated to a bank that was already busy servicing a previous request, the new request was buffered (set-aside) and waited until the previous request had completed; in the meantime, further requests to other banks could proceed unimpeded. As with caching, this approach to memory organization leads to variable access times; for this machine, the effective access time was 360ns, with a range of 180ns to 600ns. Three memory

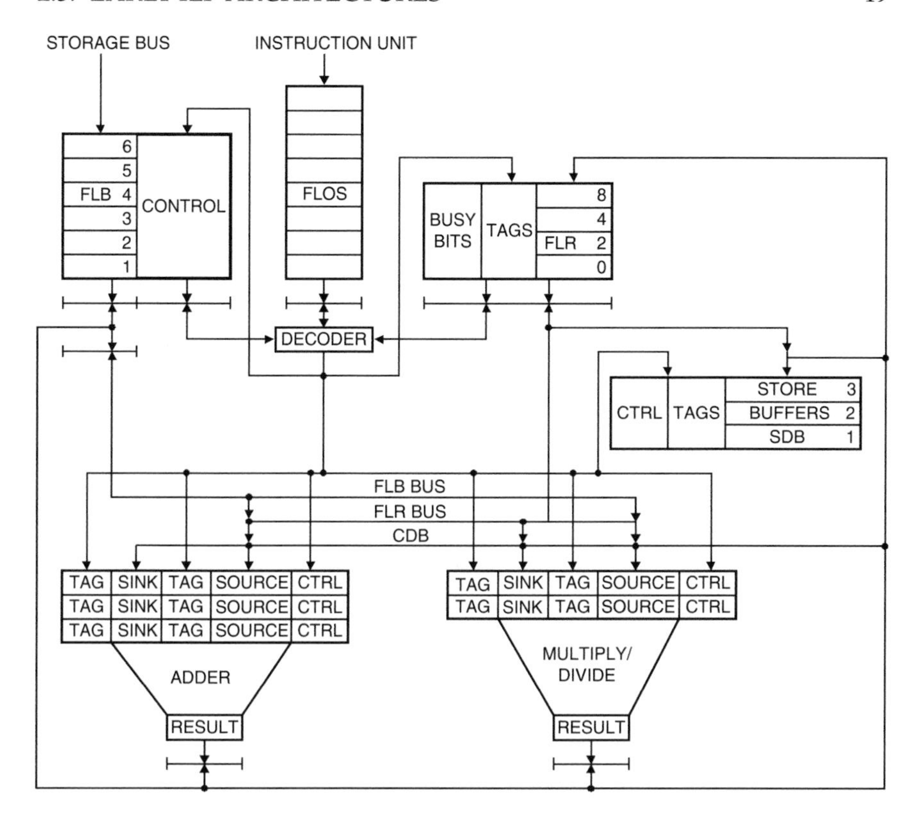

Figure 2.2 : IBM 360/91: Data Registers and Transfer Paths

address buffers and three memory data buffers provided "waiting room" for stores to memory and pending loads.

The CPU consisted of one fixed-point unit (non-pipelined, but most of the fixed-point operations executed in one 60ns cycle); a floating-point multiply/divide unit, and a floating point adder, both pipelined. Furthermore, the fetch/decode/execution of an operation was itself pipelined, with the machine fetching one new operation every cycle. The machine instruction (operation) set was, as was customary for virtually all machines of that vintage, a complex one (CISC) with most operations being able to accept one of their operands as a memory address (rather than a register). The format of the operations was of the two operand type, where the first (leftmost) operand served as both a source and destination. Fast algorithms for multiply and divide operations and carry lookahead adders were employed to provide additional concurrency. Further, the floating-point units were linked to an internal bus to support parallel dispatch of operations to the execution units.

Once fetched and decoded, operations were issued to the various units for execution. Since the actual execution of an operation, and indeed the overall fetch/decode/fetch-operands/execute/store-result cycle for each instruction, took substantially more than the 60ns cycle (pipelining) it was quite likely that preceding operations, whose results were needed by an operation being readied for issue, were not yet available. In an attempt to minimize the impact of both operation and memory access latencies and their disruption of the pipeline, and enhance chances for parallel execution, extensive buffering and lookahead techniques were used. Eight instruction-fetch buffers held prefetched instructions and effectively provided primitive caching of short loops. Processing in loop mode was also possible and the implementation was similar to that of the CDC 6600. A detailed description of the architecture is available in [AST67].

2.4 ILP Architectures in the 80's

The first true VLIW machines were mini-supercomputers in the early 1980s from three companies: Multiflow, Culler, and Cydrome. The Multiflow TRACE provided multiple functional units to facilitate parallel execution of operations; seven 32-bit opcodes were packed into a 256-bit VLIW instruction. The Culler-7, based on the Harvard architecture, was a decoupled access-execute (DAE) system from Culler Scientific Systems. The design consisted of an "A" machine to control program sequencing and data memory addressing and access, and a microcoded "X" machine for floating-point computations that could run in parallel with the A machine. The Cydra 5, developed at Cydrome, also had multiple functional units to support VLIW execution; in addition, it also had a special mode wherein operations in an instruction could be executed sequentially. Although these machines were not commercially successful, the compiler techniques that were developed were very influential. Trace scheduling and software pipelining, pioneered by Fisher and Rau, respectively, are now central pillars of VLIW compiler technology. Other early VLIW efforts include Warp [AAG+85, GKLW85, AAG+87] and CHoPP [MPS87].

Cydra 5

The Cydra 5, a heterogeneous multiprocessor, was an outcome of eight years of research at TRW Array Processors and at ESL (a subsidiary of TRW). The machine was comprised of a numeric processor and an interactive processor for all non-numeric work. The numeric processor was based on the directed-dataflow model of computation [Rau88], a successor of the polycyclic architecture developed at ESL in 1981 [RG81, RGP82]. The Motorola 68020 microprocessor was used as the interactive processor. A scheme proposed in [YYF85] was used to maintain cache coherence in the multiprocessor environment. The 64-way interleaved main memory provided high memory bandwidth. A maximum of two I/O processors could be attached to facilitate high bandwidth I/O transactions.

The architecture provided special features to enhance parallel execution of loops. The Complex Register Matrix supported overlapped execution of loop iterations; further, it could capture the complex resource usage patterns to facilitate *modulo scheduling* of loops (discussed in Chapter 6). To deal with *recurrences* (a recurrence is the use of a data value calculated in a previous loop iteration by the current loop iteration) the architecture allocated a context frame for each loop iteration. Values within a loop iteration were stored in a register context distinct from the register context of the previous iterations. Values stored within previous register contexts were accessible from the current iteration, facilitating the computation of recurrences. Further, the Conditional Scheduling Control hardware facilitated overlapped execution of loops with conditionals; the special hardware allowed conditional execution of operations based on certain predicate values.

The numeric processor of the Cydra 5 had seven functional units: a floating-point adder/integer ALU, a floating-point multiplier/divider, two memory reference ports, two address arithmetic units and a branch unit. Each functional unit was pipelined. The numeric processor supported two different instruction formats. The MultiOp instruction format had seven operations, similar to a VLIW instruction except for the existence of a predicate specifier. The typical format for each operation consisted of an opcode, two source register specifiers, one destination register specifier and a predicate register specifier. On the other hand, the UniOp instruction format allowed an issue of a single operation per instruction. The rationale behind

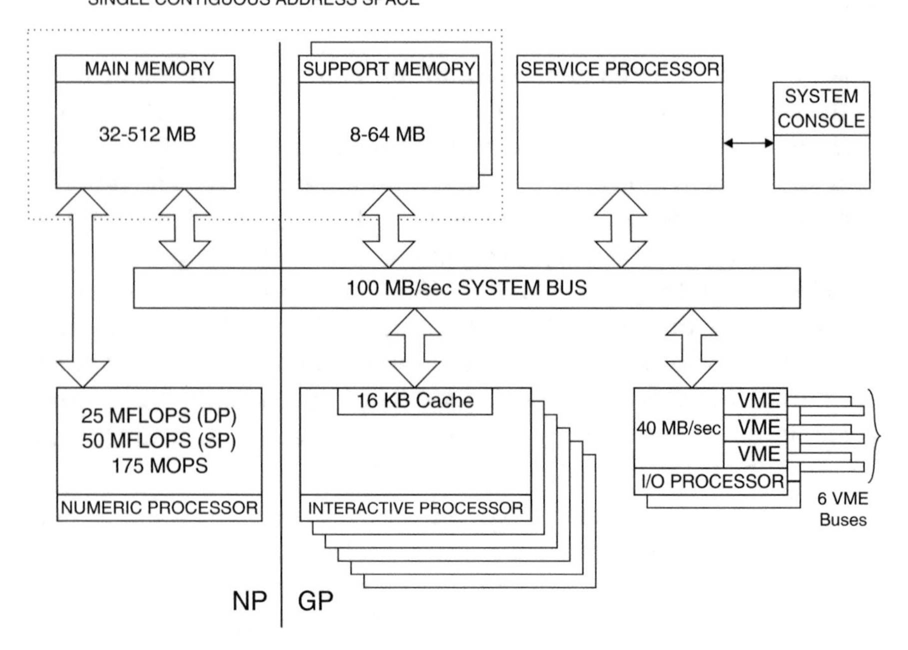

Figure 2.3 : Cydra 5 Block Diagram

the introduction of the UniOp instruction format was to mitigate the effect of MultiOp format on instruction cache performance for portions of code with little parallelism available. A detailed description of the architecture is available in [Rau88, SM88, RYYT89].

Multiflow TRACE

Multiflow Computer Inc. built the world's first VLIW computer along with—and perhaps more importantly—a compiler that could expose ILP. Multiflow showed that a compiler could find large amounts of parallelism (well beyond what was commonly believed to be the limit [TF70, FR72]) by scheduling beyond basic blocks boundaries, using a technique called *trace scheduling* (discussed in Chapter 4). There were three series of Multiflow machines, viz., the 200 series, the 300 series and the 500 series. The 200 and the 300 series came in three widths: a 7-wide, which had a 256-bit instruction; a 14-wide, which had a 512-bit instruction; a 28-wide with a 1024-bit instruction. The wider machines were realized as a combination of 7/300s. We briefly discuss the 300 series.

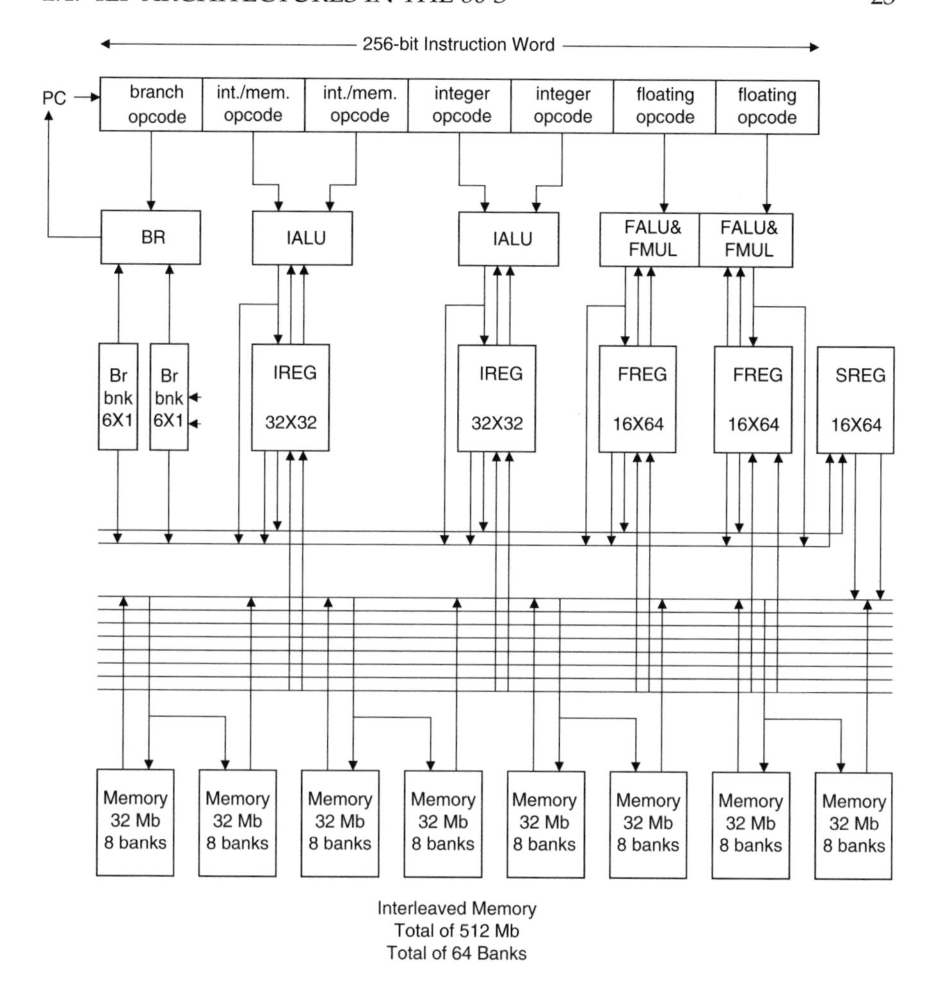

Figure 2.4 : The Multiflow TRACE 7/300 Block Diagram

The architecture of Multiflow TRACE 7/300 is shown in Figure 2.4. The core of the computation engine was built using 8000 gate CMOS gate arrays with 154 signal pins. Advanced Schottky TTL was used for "glue" logic and bus transceivers. The machine had two integer units, two floating-point units, and a branch unit. The integer units could each perform two operations per 130ns cycle (four in all), whereas a floating-point unit could perform one operation per cycle. Separate register files (sixty four 32-bit registers) were provided to the integer and floating point units; each register file could handle eight reads, writes and bus-to-bus forwards, plus bypassing from each write port to read port. The integer unit also included dynamic address translation hardware (TLB) and supported demand-paged virtual addressing. The machine had an instruction cache of 8K but had no data cache. The memory systems supported 512 MB of physical memory with up to 64-way interleaving, thus providing enormous memory bandwidth; the memory system was pipelined to exploit parallelism among memory references. It had two I/O processors which supported a 246 MB/sec DMA channel to the main memory.

Operations were RISC-like: fixed 32-bit length, fixed-format, three register operations with memory accessed through explicit loads and stores. Operations could be completed either in a single cycle or were explicitly pipelined. In the same spirit as RISC machines such as MIPS [HJP$^+$82] and the IBM 801 [Rad82], the microarchitecture was exposed to the compiler for resource management. However, TRACE provided many more functional units to facilitate exploitation of parallelism beyond basic block boundaries. An instruction could contain 4 integer operations, 2 floating-point operations and a branch operation. A detailed description of the architecture is available in [CNO$^+$88, LFK$^+$93].

2.5 ILP Architectures in the 90's

After a few years of operation, Cydrome and Multiflow both closed their doors after failing to establish a large enough market. HP hired Bob Rau and Mike Schlansker of Cydrome, and they began the FAST (Fine-grained Architecture and Software Technologies) research project at HP in 1989; this work later developed into HP's PlayDoh architecture [VK94]. In 1990, Bill Worley at HP started the PA-Wide Word project (PA-WW, also known as SWS, SuperWorkStation).

In 1992, Worley recommended that HP seek a manufacturing partner for PA-WW, and in December 1993 HP approached Intel; co-operation between the two companies was announced in June 1994. The term EPIC (Explicitly Parallel Instruction Computing) was coined to describe the design philosophy and architecture style envisioned by HP, and the instruction set architecture was named IA-64. Itanium is the name of the first implementation (it was previously known by the project name Merced) of the IA-64 [SA00]. Because Itanium exposed instruction scheduling to the compiler (in contrast to doing all instruction scheduling dynamically in hardware), we discuss the Itanium design in more detail in Section 2.6.

The late 80's and the early 90's saw a great deal of activity in the development of VLIW [KM89a, LS90], superscalar [Gro90, McG90, NNN+91] and superpipelined machines [JW89]. In the late 1980s, Intel introduced the i860 RISC microprocessor and IBM introduced the first superscalar workstation RS/6000. The i860 had two modes of operation: a scalar mode and a VLIW mode. In the VLIW mode, the processor always fetched two instructions and assumed that one was an integer instruction and the other floating-point. A single program could switch between the scalar and VLIW modes, thus implementing a crude form of code compression.

Today the VLIW philosophy is extensively used for designing processors for non-numerical applications such as the ST VLIW core for IP video telephony, the MAP1000A VLIW Mediaprocessor for media applications [BLO00]. Philips Semiconductors' VLIW TriMedia TM1300 processor chip, designed for high end applications such as video conferencing, TV set-top boxes, and digital video cameras, can issue up to 5 operations in parallel; the maximum clock frequency is 166MHz. It has 128 32-bit general purpose registers and on-chip hardware support for audio-video I/O. Similarly, the TI TMS320C6701 processor, designed for DSP applications, can execute up to 8 operations per cycle (6 float, 2 integer). It has 128K on-chip RAM and thirty two 32-bit registers [TMS]. The internal memory has two blocks: 64 KB for program memory which is configurable as cache or memory-mapped program space and 64 KB for internal data memory. The Transmeta Crusoe TM5700 and TM5900 processors provide power-efficient VLIW solutions for embedded designs. Each supports multiple integer and floating-point execution units, separate 64 KB instruction and data caches, MMU and multimedia instructions. In the rest of this section, we briefly discuss some contemporary ILP architectures.

Sun UltraSPARC

SPARC Version 9 (SPARC-V9) was the most significant change to the SPARC architecture [WG00] since its release in 1987. SPARC-V9 incorporated several important enhancements: 64-bit addresses and 64-bit integer data, features to support optimizing compilers and high performance operating systems, superscalar implementation, fault tolerance, etc. As one of the first implementations of this new architecture, the UltraSPARC-I added a number of other features like a nine-stage pipeline allowing up to 4 instructions to be issued per cycle, dynamic branch prediction, on-chip 16K data and 16K instruction caches with up to 4 MB external cache, and on-chip graphics and imaging support. It was implemented using 0.5 μm, 4-layer metal CMOS technology operating at 3.3 volts.

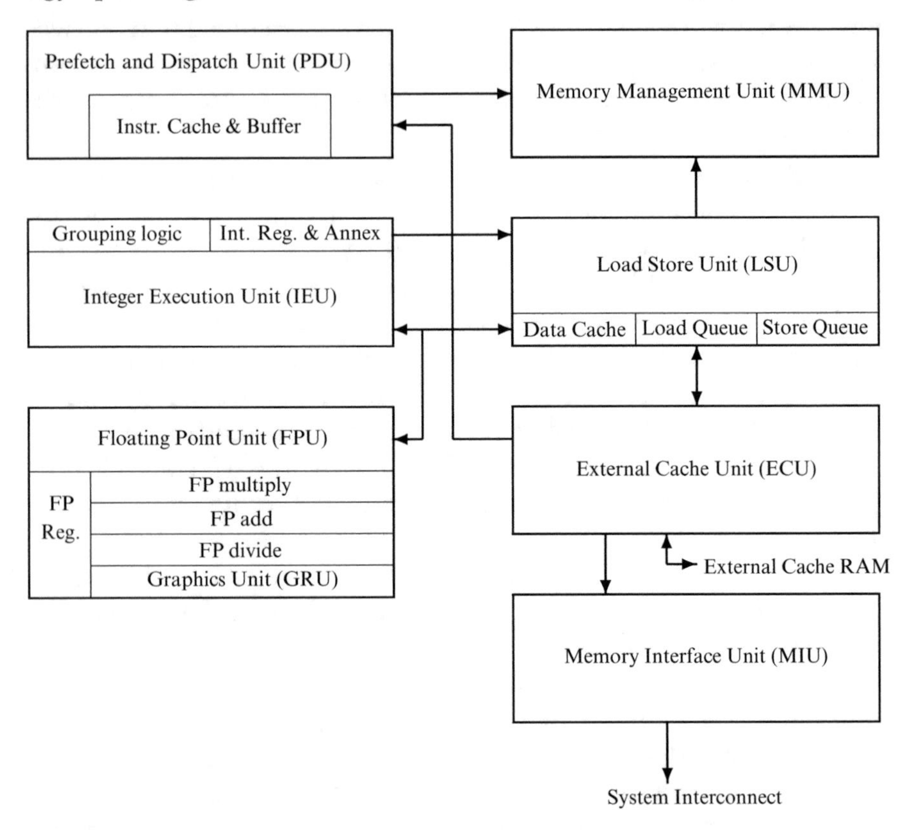

Figure 2.5 : UltraSPARC-II Block Diagram

UltraSPARC-II was the second generation member of the UltraSPARC family of processors [GBC+95, TO96]. In addition to using a new process technology, it provided a higher clock frequency, multiple SRAM modes and system to processor clock ratios. The architecture of UltraSPARC-II is shown in Figure 2.5. The Prefetch and Dispatch Unit (PDU) ensures that all execution units remain busy by pre-fetching the instructions. Instructions can be prefetched from all levels of the memory hierarchy, including the I-cache, E-cache, and main memory. The machine used a double-instruction-issue pipeline with nine stages: fetch, decode, grouping, execution, cache access, load miss, integer pipe wait, trap resolution, and writeback. The latency (time from start to end of execution) of most instructions was nine clock cycles. However, at any given time, as many as nine instructions could be executed simultaneously, producing an overall rate of execution of one clock per instruction in many cases.

The branch prediction mechanism was based on a two-bit state machine that predicts branches based on the specific branch's most recent history. A two-bit prediction was maintained for every two instructions in the I-cache, allowing state information for up to 2,048 branches to be maintained. Branch following, the ability to rapidly fetch predicted branch targets, helped reduce fetch latency by providing a fuller instruction buffer, which in turn enhanced instruction execution rate.

The Integer Execution Unit (IEU) incorporated 2 ALU's for arithmetic, logical, and shift operations; an 8-window register file; result bypassing; and a Completion Unit. The Floating Point Unit (FPU) consisted of five separate functional units to support floating-point and multimedia operations. The Memory Management Unit (MMU) provided the functionality of a reference MMU and an IOMMU, handling all memory operations as well as arbitration between data stores and memory.

Intel Pentium Pro

The Pentium Pro was a 32-bit Intel architecture microprocessor with dynamic execution features: out-of-order, speculative, superscalar, superpipelined, and micro-dataflow. It was implemented using a 0.6μm 4 layer metal BiCMOS process technology [Sch94, CS95]. High performance was achieved through the use of a large, full-speed cache accessed through a dedicated bus interface feeding a generalized dynamic execution microengine. A primary 64-bit processor bus included additional pipelining features to provide high throughput to the CPU and cache.

This processor implemented dynamic execution using an out-of-order, speculative execution engine, with register renaming of integer [HP90], floating point and flag variables, multiprocessing bus support, and carefully controlled memory access reordering. The flow of Intel architecture instructions was predicted and these instructions were decoded into micro-operations (μops), or series of μops; the μops were register-renamed, placed into an out-of-order speculative pool of pending operations, executed in dataflow order (when operands were ready), and retired to permanent machine state in source program order. This was accomplished with one general mechanism to handle unexpected asynchronous events such as mispredicted branches, instruction faults and traps, and external interrupts. Dynamic execution—the combination of branch prediction, speculation and micro-dataflow—was the key to high performance.

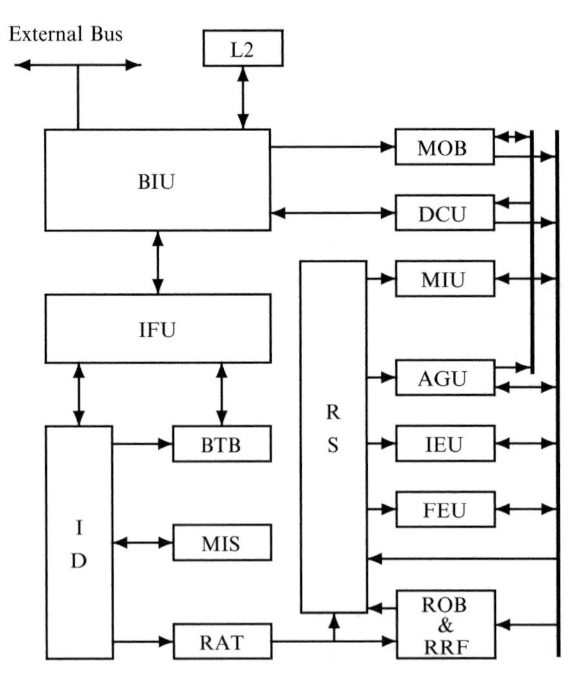

Figure 2.6 : Pentium Pro Basic CPU Block Diagram

The basic operation of the microarchitecture is described below (see Figure 2.6). The In-Order-Front-End consisted of the following parts: Instruction Fetch Unit (IFU), Branch Target Buffer (BTB), Instruction Decoder (ID), Micro-Instruction Sequencer (MIS), and Register Alias Table (RAT). The Reorder Buffer (ROB) & Real Register File (RRF) constituted the In-Order-Back-End, while the Out-Of-Order-Core consisted of the following parts: Reservation Stations (RS),

Memory Order Buffer (MOB), Data Cache Unit (DCU), Memory Interface Unit (MIU), Address Generation Unit (AGU), Integer Execution Unit (IEU) and Floating-point Execution Unit (FEU). The BIU is the Bus Interface Unit. The BTB had a non-trivial design using a 512 entry two-level adaptive algorithm. The IFU, which contained the I-cache, is where the Intel architecture instructions constituting the program lived. The BTB assisted the IFU in choosing an instruction cache line for the next instruction fetch. I-cache line fetches were pipelined with a new instruction line fetch commencing on every CPU clock cycle.

Three parallel instruction decoders (ID) converted multiple Intel Architecture instructions into multiple sets of micro-ops (μops) each clock cycle. Three micro-ops were generated per clock, making the Pentium Pro processor a superscalar processor of degree 3. The sources and destinations of these μops were renamed by the Register Alias Table (RAT), which eliminated register reuse artifacts, and forwarded to the Reservation Station (RS) and to the ReOrder Buffer (ROB). The renamed μops were queued in the RS where they waited for their source data, which could come from several places, including immediates, data bypassed from just-executed μops, data present in a ROB entry, and data residing in architectural registers (such as EAX). The queued μops were dynamically executed according to their true data dependences and execution unit availability (IEU, FEU, AGU). The order in which μops executed had no particular relationship to the order implied by the source program. Memory operations were dispatched from the RS to the Address Generation Unit (AGU) and to the Memory Ordering Buffer (MOB). The MOB ensured that memory access ordering rules were observed.

Once a μop had executed and its destination data was produced, that result datum was forwarded to subsequent μops needing it, and the μop became a candidate for *retirement*, the act of taking the machine's speculative state and irrevocably committing it to permanent machine state. Retirement hardware in the ROB used μop timestamps to reimpose the original program order on the μops as their results were committed to permanent machine state in the Retirement Register File (RRF). The retirement process observed not only the original program order, it also correctly handled interrupts and faults, and flushed all or part of its state on detection of a mispredicted branch. When a μop was retired, the ROB wrote that μop's result into the appropriate RRF entry and notified the RAT of that retirement so that subsequent register renaming could be activated. The component included separate data and instruction L1 caches (each 8KB), and a unified 256KB non-blocking L2 cache. The L1 Data Cache was dual-ported and non-blocking, supporting one load and one store per cycle. The L2 cache interface ran at the full CPU clock speed, and could transfer 64 bits per cycle. The external bus was also 64-bit and could sustain a data transfer every bus-cycle. This external bus operated at 1/2, 1/3, or 1/4 of the CPU clock speed. Further discussion of the Intel Pentium Pro is available in [Pap96].

Mips R10000

The Mips R10000 was a four-way superscalar RISC processor implementing the 64-bit Mips 4 instruction set architecture. It was implemented using 0.35-μm CMOS process technology on a 16.64×17.934 mm chip. The processor had five independent pipelined execution units (see Figure 2.7), including a non-blocking load/store unit, dual 64-bit integer ALU's, and 64-bit IEEE standard 754-1985 floating point units. In addition, there was a floating point adder and a multiplier,

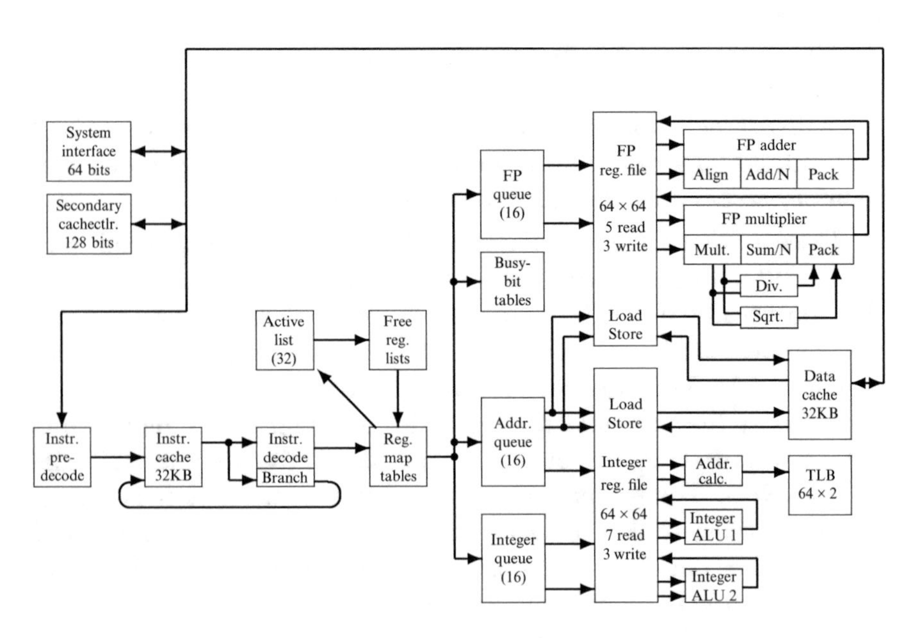

Figure 2.7 : Mips R10000 Block Diagram

each with two-cycle latency. The hierarchical, non-blocking memory subsystem included:

- two on-chip, two-way associative primary caches: a 32KB instruction cache and a 32KB two-way-interleaved data cache,
- an external two-way-associative secondary cache, a 128-bit wide synchronous static RAM, and
- a 64-bit multiprocessor system interface with split transaction protocol.

The R10000 fetched and decoded four instructions per cycle, and speculatively executed beyond branches with a four-entry branch stack. The R10000 implemented non-blocking caches to overlap cache refill operations (executed on a miss in the instruction cache) with other instructions that could continue out of order. The processor looked ahead up to 32 instructions to find parallelism. It predicted the direction of conditional branches and fetched instructions speculatively along the predicted path. The prediction was made using a 2-bit algorithm based on a 512-entry branch history table. In the Mips architecture, like the early RISC processors, a delay slot instruction, i.e., the instruction immediately following a branch or a jump, was executed while reading the target instruction from the cache. When a program took a jump or branch, the processor discarded the instructions fetched beyond the delay slot.

There were six independent pipelines: an instruction fetch pipeline and a five stage execution pipeline corresponding to each pipelined functional unit. The instruction fetch pipeline occupied stages 1–3 of the R10000 core pipeline. In stage 1, R10000 fetched and aligned the next four instructions. In stage 2, it decoded and renamed these instructions using map tables and computed the target addresses for the branch and jump instructions. In stage 3, it wrote the renamed instructions into instruction queues; the integer and floating-point instructions had separate queues. Instructions waited in these queues until their operands were ready. The processor read operands from the register files in the second half of stage 3 and issued the instruction for execution in stage 4. The result of an execution was either written into a register file or bypassed into operand registers; there are no separate structures such as reservation stations or reorder buffers. A detailed description of the R10000 architecture is available in [Yea96].

IBM/Motorola/Apple PowerPC

In February of 1990 IBM started the delivery of RS/6000, an implementations of the POWER Architecture [BGM90]. The POWER architecture embodied most of the RISC features: instructions were of fixed length (4 bytes), with consistent formats, and the architecture was load-store. The architecture provided a set of general purpose registers for fixed-point computation, including the computation of memory addresses; in addition, it provided a separate set of floating-point registers for floating-point computation. In October of 1993 IBM delivered the POWER2 processor [WD94] which had twice the number of execution units of the earlier POWER designs.

PowerPC Architecture was a 64-bit architecture, a superset of the POWER 32-bit architecture. This architecture extended addressing and fixed-point computation to 64 bits, and supported dynamic switching between the 64-bit mode and the 32-bit mode. The first PowerPC processor, the MPC601, was shipped in 1994; later, Motorola (now Freescale Semiconductor Inc.) and IBM unveiled the MPC604 processor with performance upto 100 MHz. In 2000, Freescale introduced the MPC74xx family of PowerPC processors (G4 family). The G4 core called the MPC7447A processor is a 32-bit implementation of the superscalar PowerPC RISC architecture, operates at 1.42 GHz and has on-chip power management features. The key architectural features include 512 KB of on-chip L2 cache, a 64-bit bus interface and a full 128-bit implementation of Freescale's AltiVec™ technology [Ful98]. The AltiVec technology features SIMD (Single Instruction, Multiple Data) functionality that supplements the host processor with an advanced 128-bit vector execution unit specifically designed to accelerate embedded application processing needs. This new engine facilitates simultaneous execution of up to 16 operations in a single clock cycle.

The block diagram of the MPC7447A PowerPC processor is shown in Figure 2.8. The processor has 11 independent execution units: four integer units (three simple plus one complex), a double-precision floating point unit, four AltiVec technology units (simple, complex,

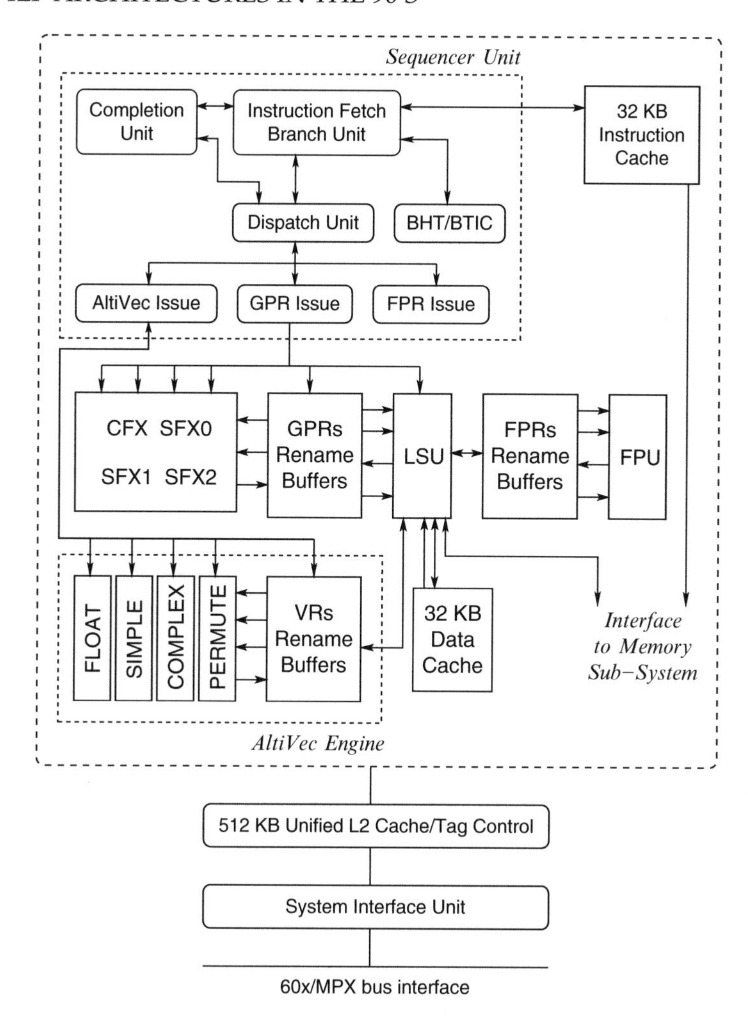

Figure 2.8 : The MPC7447A PowerPC processor

floating and permute), a load/store unit and a branch processing unit. The processor is capable of issuing four instructions per clock cycle (three instructions plus one branch). The microprocessor reference manual has further details [MPC05].

In June 2003, Apple introduced the superscalar PowerPC G5 which supported up to 215 in-flight instructions. It was implemented using 90 nm, silicon-on-insulator (SOI) technology. The processor had 12 functional units, two double-precision floating-point units, and a 128-bit Velocity Engine for image and video processing applications.

In addition, it had fast L1 and L2 caches and a dual-channel frontside bus at up to 1.25 GHz which provided high bandwidth to and from the rest of the system, allowing large numbers of tasks to run concurrently. Further, it supported group instruction dispatching, deep queues, and three-stage branch prediction logic. The success or failure of each prediction was captured in three large 16K branch history tables local, global, and selector that were used to improve the accuracy of future branch predictions. Also, the PowerPC G5 supported symmetric multiprocessing; the dual independent frontside buses allow each processor to handle its own tasks at maximum speed with minimal interruption.

HP PA-8000

The PA-8000 RISC CPU is the first of a new generation of Hewlett-Packard microprocessors. It was implemented using HP's 0.5-μm, 3.3-V CMOS process; the die measures 17.68 mm$\times$19.1 mm and contains 3.8 million transistors. It is a four-way superscalar machine that supports 64-bit computing and combines speculative execution with dynamic instruction reordering. The PA-8000 provides glue logic support for up to four-way multiprocessing via a Runway system bus, a 768-MB/s split-transaction bus that allows each processor to generate multiple outstanding memory requests. The PA-8000 implements PA (Precision Architecture) 2.0, a binary compatible extension of the previous PA-RISC architecture.

The functional block diagram of PA-8000 is shown in Figure 2.9. The functional units in the processor include two 64-bit integer ALU's, two 64-bit shift/merge units, dual floating-point multiply-and -accumulate (FMAC) units, and dual divide/square root units. Each FMAC unit is optimized to perform the operation $A * B + C$. The FMAC units are fully pipelined, while the divide/square root units are not. The peak throughput of the PA-8000 is four floating-point operations per cycle [DH96]. The processor incorporates two complete load/store pipes, including two address adders, a 96-entry dual-ported TLB, and a dual-ported cache. The dual load/store units and the memory system interface are shown on the right side of Figure 2.9.

The most important feature of the architecture is the 56-entry instruction reorder buffer, consisting of the ALU buffer that can store

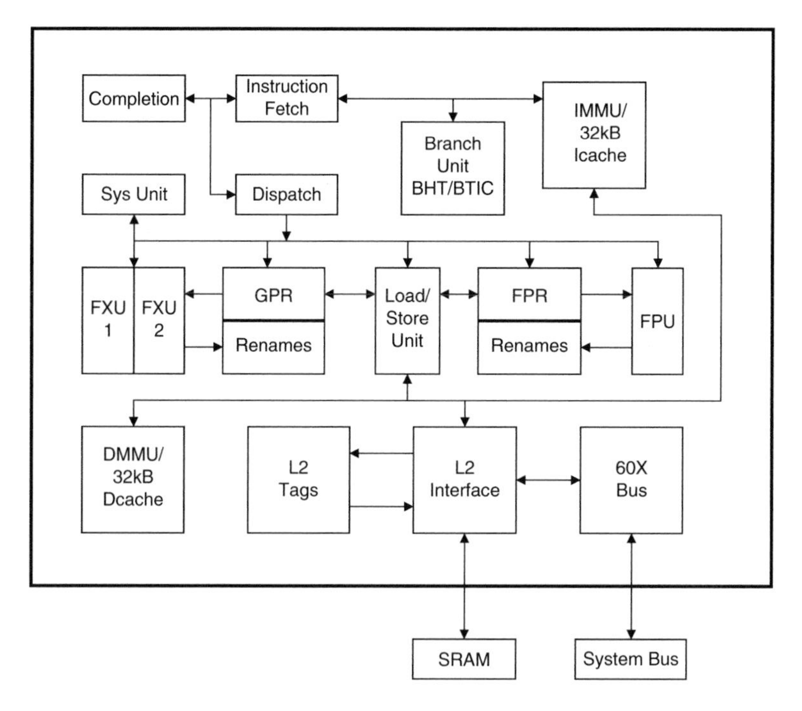

Figure 2.9 : Block Diagram of the HP PA-8000

28 computation instructions and the memory buffer that can store 28 load and store instructions. These buffers track the dependences between the instructions they hold and let any instructions in the window execute as soon as they are ready to fire. The reorder buffer also tracks branch prediction outcomes. The PA-8000 has a 32-entry branch target address cache (BTAC). The BTAC is fully associative; it associates the address of a branch instruction with the address of its target. The processor implements both static and dynamic modes of branch prediction. Each TLB entry contains a bit that indicates which mode should be used by the branch prediction hardware. The software can select the mode on a page-by-page basis. A detailed description of the HP PA-8000 architecture is available in [Kum97].

2.6 Itanium

Intel and Hewlett-Packard teamed up in the late 1990s to develop a new 64-bit processor design. The goal was to design an architecture with explicit support for parallel execution and its instruction set. In addition, the designers wanted to handle branches efficiently. These goals eventually led the design of the *Explicitly Parallel Instruction Computing (EPIC)* architecture [SR00]. In the rest of the section, we first discuss the EPIC philosophy and then present a brief overview of the Itanium architecture.

2.6.1 The EPIC Philosophy

The EPIC architecture represented an evolution of VLIW that incorporated many superscalar concepts; it retains the VLIW philosophy of parallelizing the original sequential code at compile time, but augments it with hardware features—akin to those in superscalar processors—to better cope up with what actually happens at runtime. In comparison with superscalar architectures, EPIC provides an alternate style of architecture with support for high levels of ILP with reduced hardware complexity. Further, EPIC provides architectural features for compiler-driven control of mechanisms that the microarchitecture usually handles, such as cache hierarchy management and the associated decisions as to what data to promote up the hierarchy and what to replace.

EPIC inherits the architectural ideas pioneered at Multiflow (hardware support for control speculation of loads, parallel issue of multiple branch operations) and Cydrome (predicated execution, support for software pipelining in the form of rotating register files and special loop-closing branches). The key innovations in the EPIC architecture were:

- Inclusion of wired-OR and wired-AND predicate setting compares and the availability of two target compares.
- Branch support for software pipelining of while loops.
- Architectural support for data speculation.
- Support for data cache bypassing, specifically inclusion of the source specifier and the target cache specifier to deal with the data cache hierarchy.

Because most codes (especially non-floating point codes) have frequent conditional branches, any limitations in scheduling branch operations implies that branch operations will be a performance bottleneck in ILP. EPIC provides architectural features to better overlap branch operations with other operations. Branch operations in EPIC are divided into three components: i) prepare-to-branch, which computes the branch's address, ii) a compare, which computes the branch condition, and iii) an actual branch, which specifies when control is transferred. Prescheduling of prepare-to-branch and compare operations enables speculative prefetching of instructions at the branch target. After executing the compare operation, the hardware dismisses the unnecessary speculative prefetches. These mechanisms allow parts of a branch computation to be scheduled in parallel with other operations without relying on the movement of the actual branching operation itself.

The ever widening processor-memory performance gap has been the driving force to use caches to hide memory latency. EPIC provides architectural features to facilitate compiler-driven channeling of data through the cache hierarchy, including load operations with a source cache specifier that the compiler uses to dictate to the hardware where within the cache hierarchy it can expect to find the data. The prediction can be done based on analytical or cache miss profiling techniques. In a similar fashion, the compiler uses the target cache specifier to indicate the cache level to which the load or store should promote or demote the data. The target cache specifier helps reduce cache misses in the first- and second-level caches by controlling their contents. The compiler can exclude data with low temporal locality from the first- or second-level cache, and can remove data from the cache level on the last use. Further, the data prefetch cache allows the prefetching of data with poor temporal locality into a low-latency cache without displacing the first-level cache's contents.

Static aliasing among memory references poses another impediment to exploiting parallelism. In general, a compiler must be conservative, i.e., it assumes aliasing between two memory references unless it can prove that they cannot alias. In order to alleviate this limitation, EPIC provides support for data speculation, whereby a conventional load operation is broken down into two components: data-speculative load and data-verifying load. The compiler moves a data-speculative load above potentially aliasing stores to hide the memory access latency. It schedules the data-verifying load after the potentially aliasing stores and uses hardware to detect whether the memory locations are in fact the same (aliased). In the absence of aliasing, execution proceeds as usual; otherwise, the processor is stalled and the load operation is re-executed. Further, EPIC provides support for aggressive data-speculative code motion, wherein operations that use the result of data-speculative loads are also moved above potentially aliasing stores. Next, we discuss the Itanium architecture, an example of the EPIC philosophy.

2.6.2 Itanium Architecture

The Intel Itanium processor, a 64-bit architecture, was the first implementation of the family of processors based on the IA-64 architecture. The processor employs EPIC design concepts (discussed in the previous subsection) for exploiting higher levels of ILP; it goes beyond RISC and CISC approaches by coupling hardware with compiler technology to make parallelism explicit to the processor. The hardware provides abundant execution resources and supports various dynamic run-time optimizations for high performance.

The (first generation) Itanium processor provides a 6 instruction-wide and 10-stage deep pipeline, running at 800 MHz on a 0.18-μm process. This provides both large hardware support to exploit ILP as well as high frequency for minimizing the latency of each instruction. The processor provides the following execution units: 4 integer units, 2 floating point units, 4 multimedia units, 2 load/store units and 3 branch units. Performance beyond compile-time parallelization is achieved via dynamic prefetch, branch prediction, a register scoreboard and non-blocking three-level caches. Further, Itanium supports an SIMD instruction set architecture (ISA), facilitating exploitation of data parallelism; the SIMD ISA is compatible with Intel's MMX technology. Similarly, it supports floating point SIMD operations and is compatible with the Intel's Streaming SIMD extensions (SSE). The basic block diagram of the microarchitecture is shown in Figure 2.10.

Itanium provides a large number of both integer and floating point registers: 128 64-bit integer registers, 128 82-bit floating point registers, 64 predicate registers and 8 branch registers for speculative execution. Registers are renamed to facilitate function calls via variable-size windows; the registers are rotated in the context of software pipelining of loops. Further, Itanium provides hardware support for floating point MAC operations, maximum and minimum operations, and fully pipelined divide and sqrt primitives. It also supports an IA32 hardware execution unit for backward compatibility with

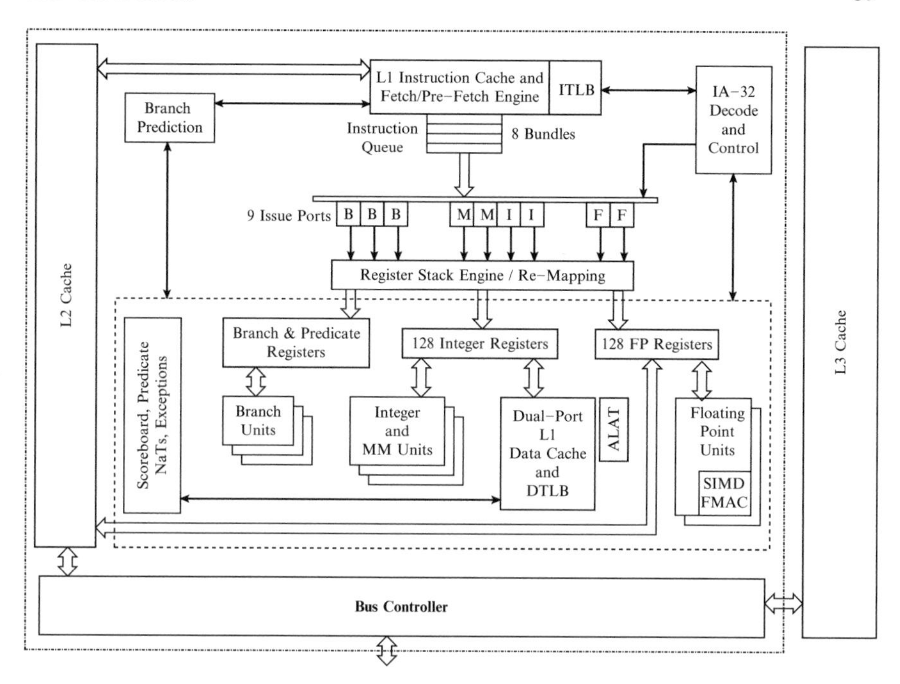

Figure 2.10 : Itanium Processor Block Diagram

traditional Intel x86 instructions. In addition, Itanium provides hardware support, called ALAT (Advanced Load Address Table), for data speculation; ALAT contains information about outstanding advanced loads. The register stack engine spills/fills registers as needed and works upon stack overflow/underflow; it ensures efficient stack register usage and provides a programming model that provides a register stack with a logically infinite number of registers.

The Itanium processor has three levels of processor cache. The instruction and data L1 caches are 16KB in size, 4-way set associative with 32B line size. The L1 data cache is dual-ported; thus, it can support two concurrent loads/stores. The L1 data cache only caches data for the integer unit, not for the floating-point units. The unified L2 cache is 96KB in size, 6-way set associative with a 64B line size and uses a write-back with a write-allocate policy. The L3 cache is either 2MB or 4MB in size, 4-way set associative with 64B line size. The L3 cache is accessed via a high bandwidth 128-bit back-side bus allowing memory accesses at processor core speed.

Itanium executes instructions in parallel in what are known as *instruction bundles*, comprised of 3 instructions each. The processor attempts to schedule two instruction bundles in parallel per clock, allowing the processor to schedule and execute a maximum of 6 instructions in a single clock cycle. Each 128-bit sized bundle contains 41-bit instruction slots and a 5-bit template field, as shown in Figure 2.11.

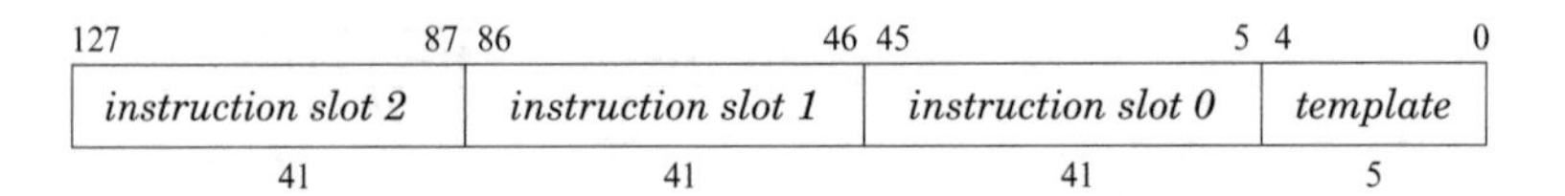

Figure 2.11 : Bundle Format in Itanium

The core pipeline of Itanium consists of 10 stages, as shown in Figure 2.12. First, the IPG stage generates the instruction pointers. Then the FET stage fetches instructions by accessing L1 I-Cache and L1-ITLB. The ROT stage formats the instruction stream and loads the instruction buffer to be used by the rest of the pipeline.

IPG	FET	ROT	EXP	REN	WLD	REG	EXE	DET	WRB

Figure 2.12 : 10-stage Itanium pipeline

The EXP stage expands the instruction templates and issues the instructions. The REN stage handles register renaming for register stack and register rotation. The WLD (word-line decode) stage decodes the instructions and the REG stage delivers data to the functional units either from the registers or by bypassing the data generated by functional units for consumption by chained instructions. This stage also generates spill or fill instructions required by the register stack engine. The EXE stage delivers instructions and the data to the functional units. The DET stage detects exceptions and branch mispredictions. The pipeline is flushed by exceptions and branch mispredictions, causing highest priority pipeline stalls. Stalls in data delivery and floating point micropipelines are also detected at this stage. Finally, the WRB stage writes the output to the appropriate output registers.

The Itanium architecture provides explicit support for software pipelining, discussed in Chapter 7. The pipeline stages are defined using full predication and other special branch handling features specific to software pipelining (e.g., loop branches and loop registers LC, EC). Register rotation is used to alleviate loop copy overhead; similarly, predicate rotation is used to remove the overhead associated with the prologue and epilogue. Further, the hardware supports multi-way branching, a key requirement for software pipelining of loops with conditionals.

FURTHER READING

The history of parallelism is very well discussed in the book by Hockney and Jesshope [HJ81]. A detailed discussion on the evolution of computer architecture is available in [Bae80, SBN82, Smi88]. As discussed earlier, the development of CDC 6600 and IBM 360 mark the onset of ILP processors; further discussion on the CDC 6600 and the IBM 360 family is available in [AWZ64, CDF64] and [Amd64, ABJ64] respectively. The state of the art for minis in the mid-seventies is well covered in [Kou73]. The 4- and 8-bit microprocessors are well-covered in [HJ76].

An early empirical evaluation of some features of instruction set processor architectures is presented in [Lun75, Lun77]. The RISC concept was first introduced in [PS81, PS82], which was later enhanced in [KSPS83]. A comprehensive discussion of the architecture and the pipeline of RISC 1 and RISC II is available in [Kat85]. One of the early RISC architectures, SOAR, is discussed in [UBF+84]. A discussion of RISC principles, with good explanation of instruction pipelines and how RISC can take advantage of them is present in [Pat85].

Overlapped execution in microprogramming was discussed in [Hig78]; an excellent discussion of overlap and pipelining can be found in [CK75, RL77]. Support for overlapped loop execution is discussed in [DHB89]. A detailed description of the IBM RISC System/6000 is discussed in [BGM90]. The book by Stone [Sto93] is a good reference for computational scientists, containing detailed explanations of pipelining and memory organization, chapters on scientific applications, vector machines, and parallel processing. The book by Trew [TW91] presents a good survey of parallel computer systems.

Look-ahead processors, the forerunners of today's superscalar processors built in the 1960s, performed out-of-order execution while issuing multiple operations per cycle. A survey of the early lookahead processors is available in [Kel75]. An excellent reference on superscalar processor design and its complexity is the book by Johnson [Joh91]. On the other side of the spectrum, the earliest VLIW processors built were the so-called attached array processors [RC69, IBM76, FSP79, INT89]. Some of the popular products were the Floating Point Systems AP-120B, the FPS-164 and the FPS-264. A very good reference for processor arrays is the book by Hockney and Jesshope [HJ88]; the book is also a good resource for computational scientists, with a nice history of high performance computing and a comprehensive survey of parallel algorithms for important matrix operations in addition to parallel and vector computer architecture. The book by Hwang [Hwa93] is a good reference for facts about machines such as, but not limited to, CM-5, KSR-1, and Paragon X/PS. The reader is advised to refer to [FO84, Fis87, Fis90, BYA93] for further discussion of the early VLIW architectures. The performance of Intel i860 microprocessor is discussed in [Atk91]. A detailed description of the Warp architecture is available in [GO98]. A VLIW approach towards embedded computing is discussed in [FFY04].

3

SCHEDULING BASIC BLOCKS

A basic block in a program is a sequence of consecutive operations, such that control flow enters at the beginning and leaves at the end without internal branches. While basic block scheduling is the simplest non-trivial instruction scheduling problem, it is also the most fundamental and widely used in both software and hardware implementations of instruction scheduling. This chapter introduces basic terminology used in all subsequent chapters and covers a number of different approaches to basic block scheduling.

3.1 Introduction

A *basic block* in a program is a sequence of consecutive operations, such that control flow enters at the beginning and leaves at the end without internal branches. There are no jumps into, within, or out of a basic block, though we may allow a branch statement as the last statement in the block provided it cannot branch back into the middle of the block. *Basic block parallelization* overlaps the execution of operations in a basic block without changing the result that would have been obtained by executing the basic block in the original sequential order.

The sequential execution order imposes a *dependence structure* on the set of operations, determined by how they access different memory locations. A new order of execution is *valid* if whenever an operation B *depends* on an operation A in the block, execution of B in

A. Aiken et al., *Instruction Level Parallelism*,
DOI 10.1007/978-1-4899-7797-7_3

the new order does not start until after the execution of A has ended. The basic assumption is that executing the operations in any valid order cannot change the final results expected from the basic block.

To reduce the total execution time of the block, one needs to find a new valid order where operations are overlapped. Furthermore, among all the valid orders, one must choose from those that are compatible with the physical resources of the given machine. Even when the simultaneous processing of two or more operations is permissible by dependence considerations alone, there may not be enough resources available to process them simultaneously.

There are many algorithms for basic block parallelization. We present four in detail: two for a hypothetical machine with unlimited resources and two for a machine with limited resources. We start with a section on basic concepts, and after developing the algorithms, end with a simple example that compares the actions of all four on a given basic block. Short descriptions of more algorithms are given in the final section.

3.2 Basic Concepts

By an *operation* we mean any atomic operation that a machine can perform. An *assignment operation* reads one or more storage locations (which may be registers or memory) and writes one location. It has the general form:

$$A\colon\ x = E$$

where A is a label for the operation, x a variable, and E an expression. Such an operation reads the memory locations specified by E and writes the location x. A *basic block* contains only assignment operations, except for possibly the last operation which may be either a conditional or unconditional branch. Flow of control always enters at the top of a basic block and always leaves at the bottom. There are no entry points except at the beginning and no branches, except possibly at the end. Consider a basic block of n operations, and let $\mathcal{B}$ be the set of operations in the block. There is a mapping $c : \mathcal{B} \to \{0, 1, 2, \dots\}$ that gives the cycle times of the operations.

Definition 3.1. *An operation B in the block* depends *on another operation A, written $A\,\delta\,B$, if A is executed before B in the original sequential order and one of the following holds:*

1. *B reads the memory location written by A;*
2. *B writes a location read by A;*
3. *A and B both write the same location.*

Case (1) is called a *true dependence*, because B actually consumes the value produced by A and there is no way to modify A or B to remove the dependence and allow the two operations to execute in parallel. Case (2) is called an *anti-dependence* (because it reverses the order of the read and the write) and case (3) is called an *output dependence* (because both operations write their output to the same location). Together, cases (2) and (3) are called *false dependences*. False dependences can often be removed or at least mitigated by simple changes to the program. For example, in case (2) we could change the location l that B writes to a new, temporary location l', which would make A and B *independent* (i.e., neither depends on the other) and allow them to execute in parallel. A new operation C that copies l' to l is added immediately after B. This copy operation still has a dependence with A but can potentially be overlapped with other operations.

While some false dependences can be completely removed, usually some false dependences are necessary and cannot be entirely eliminated from the program. Unless otherwise stated, throughout the book we assume that the program is fixed (i.e., whatever false dependences are going to be removed have already been removed) and we refer to dependences in the program to mean any of true, anti- or output dependences. Our examples almost exclusively use true dependences, however. An in-depth treatment of dependences and how they are discovered is given in [Ban97].

An operation B is *indirectly dependent* on an operation A if there is a transitive sequence of (one or more) dependences beginning at A and ending at B. Formally, we write $A\,\bar{\delta}\,B$ if there exists a sequence of operations $A_1, A_2, \ldots, A_k$, such that

$$A = A_1, A_1\,\delta\,A_2, \ldots, A_{k-1}\,\delta\,A_k, A_k = B.$$

Two operations A and B are *mutually independent* if $A\,\bar{\delta}\,B$ and $B\,\bar{\delta}\,A$ are both false. The *dependence graph* of the basic block is a directed acyclic graph, such that the nodes correspond to the operations, and there is an edge from a node A to a node B if and only if $A \,\delta\, B$. Thus, $A \,\bar{\delta}\, B$ holds if and only if there is a directed path from the node A to the node B in the dependence graph.

An execution order O for the operations in $\mathcal{B}$ is *valid* if for every A, B in $\mathcal{B}$, if $A \,\delta\, B$ then A completes before B begins in O. Because valid orders preserve all dependences, they also preserve the final results of the original execution order. The goal is to find a valid execution order that overlaps operations as much as possible. However, any such order must also be compatible with the resources of the given machine.

Control steps for execution of the basic block are numbered $1, 2, 3, \ldots$. *Scheduling* an operation in the block means assigning a control step for the start of the operation's execution. An *instruction* for a given machine is a set of operations that the machine can perform simultaneously. An instruction may be empty. *Scheduling* the basic block means creating a sequence of m instructions $(\mathbf{I}_1, \mathbf{I}_2, \ldots, \mathbf{I}_m)$, where $\mathbf{I}_k$ starts in control step k, such that

1. Each instruction consists of operations in $\mathcal{B}$, and each operation in $\mathcal{B}$ appears in exactly one instruction;

2. The operations in each instruction are pairwise mutually independent;

3. If an operation B in an instruction $\mathbf{I}_k$ depends on an operation A in an instruction $\mathbf{I}_j$, then $j + c(A) \leq k$;

4. In any control step, the machine has enough resources to process simultaneously all operations being executed.

Such a sequence of instructions is a *schedule* for the basic block. A schedule for the block can be specified indirectly by scheduling each individual operation, with all operations starting in control step k constitute the instruction $\mathbf{I}_k$.

The *weight* of a path $(A_1, A_2, \ldots, A_k)$ in the dependence graph for $\mathcal{B}$ is the sum of the operation latencies $[c(A_1) + c(A_2) + \cdots + c(A_k)]$. A path is *critical* if it has the greatest possible weight among all paths in the graph. Let T_0 denote the weight of a critical path. Then, any schedule for $\mathcal{B}$ will need at least T_0 cycles to finish.

For each operation $A \in \mathcal{B}$, the set of all immediate predecessors (in the dependence graph) is denoted by $\mathrm{Pred}(A)$ and the set of all immediate successors by $\mathrm{Succ}(A)$:

$$\mathrm{Pred}(A) = \{B \in \mathcal{B} : B\,\delta\,A\}, \quad \mathrm{Succ}(A) = \{B \in \mathcal{B} : A\,\delta\,B\}.$$

The number of members of a set S is $|S|$.

3.3 Unlimited Resources

In this section, we assume that the given machine has an unlimited supply of resources (functional units, registers, etc.). Consequently, operations in the basic block can be scheduled subject only to the dependence constraints between them; i.e., we drop the fourth requirement in the definition of a schedule given in the previous section. The two algorithms considered in this section complete the execution of $\mathcal{B}$ in T_0 cycles, where T_0 is the length of the critical path.

The ASAP algorithm schedules an operation *as soon as possible* so that the basic block can be processed in the shortest possible time. It creates a function $\ell : \mathcal{B} \to \{1, 2, \ldots\}$ such that $\ell(A)$ is the earliest possible step when A can start executing. The ALAP algorithm schedules an operation *as late as possible* within the constraint of executing $\mathcal{B}$ in the shortest possible time. It creates a function $L : \mathcal{B} \to \{1, 2, \ldots\}$ such that $L(A)$ is the latest possible step when A can start executing. The *ASAP label* of A is $\ell(A)$ and its *ALAP label* is $L(A)$. It is clear that $\ell(A) \leq L(A)$ for each operation A. It follows that any other schedule must choose a control step for A in the range $\{\ell(A), \ell(A) + 1, \ldots, L(A)\}$.

3.3.1 ASAP Algorithm

The goal of the ASAP algorithm (Figure 3.1) is to compute the ASAP label ℓ for each operation in the given basic block $\mathcal{B}$. If A and B are two operations such that $A\,\delta\,B$, then after starting A, one must wait at least until A has finished before starting B. Thus, every operation B must satisfy for every predecessor A of B the scheduling constraint

$$\ell(B) \geq \ell(A) + c(A)$$

Algorithm 3.1. Given a basic block $\mathcal{B}$, its dependence graph, and a machine with unlimited resources, this algorithm computes the ASAP label ℓ of each operation, and the total number of instructions, m, needed to replace $\mathcal{B}$. For each $A \in \mathcal{B}$, the sets $\mathrm{Pred}(A)$ and $\mathrm{Succ}(A)$ are assumed to be known.

$m \leftarrow 1$
for each operation $A \in \mathcal{B}$ **do**
  $\ell(A) \leftarrow 1$
endfor
do
  for each operation $A \in \mathcal{B}$ **do**
    for each $S \in \mathrm{Succ}(A)$ **do**
      $\ell(S) \leftarrow \max\{\ell(S), \ell(A) + c(A)\}$
      $m \leftarrow \max\{m, \ell(S)\}$
    endfor
  endfor
until there are no changes to $\ell(\cdot)$

Figure 3.1 : The ASAP Algorithm

Initially, all operations are assigned to control step 1, that is, $\ell(A)$ is initialized to 1 for all $A \in \mathcal{B}$. The basic step of the algorithm is to pick an operation B and check whether some predecessor A of B violates the scheduling constraint because $\ell(B) < \ell(A) + c(A)$. If so, $\ell(B)$ is updated to be equal to $\ell(A) + c(A)$—i.e., B is moved to the earliest instruction that satisfies the dependence on A. Operations are repeatedly rescheduled in this fashion until all operations satisfy the scheduling constraint.

Note that initially the schedule is as short as possible, as there must always be at least one instruction in any non-empty program. Because the algorithm only delays an operation when it is necessary to satisfy a dependence, it follows that at each step the current schedule length is always less than or equal to the minimum possible schedule length. Thus, when the algorithm terminates with a schedule satisfying all dependences, the schedule will also be a shortest possible legal schedule. It is also easy to see that the earliest control step where an operation B can start is given by

$$\ell(B) = \max_{A \in \mathrm{Pred}(B)} [\ell(A) + c(A)].$$

and that the total number of cycles needed to execute the block is

$$\max_{A \in \mathcal{B}} [\ell(A) + c(A)] - 1$$

which is equal to the weight T_0 of a critical path in the dependence graph. The total number of instructions needed to replace the basic block is given by $m = \max_{A \in \mathcal{B}} \ell(A)$.

Algorithm 3.2. Given a basic block $\mathcal{B}$, its dependence graph, and a machine with unlimited resources, this algorithm computes the ALAP label L of each operation, and the total number m of instructions needed to replace $\mathcal{B}$. For each $A \in \mathcal{B}$, the sets $\mathrm{Pred}(A)$ and $\mathrm{Succ}(A)$ are assumed to be known.

```
m ← 0
for each operation A ∈ B do
   L(A) ← −c(A)
   m ← min{m, L(A)}
endfor
do
   for each operation A ∈ B do
      for each operation P ∈ Pred(A) do
         L(P) ← min{L(P), L(A) − c(P)}
         m ← min{m, L(P)}
      endfor
   endfor
until there are no changes to L(·)
m ← |m|
for each operation A ∈ B do
   L(A) = L(A) + m + 1
endfor
```

Figure 3.2 : The ALAP Algorithm

3.3.2 ALAP Algorithm

The goal of the ALAP algorithm (Figure 3.2) is to compute the ALAP label L for each operation in the given basic block $\mathcal{B}$. The entire block must be completed in the shortest possible time in such a way that each operation starts as late as possible. The basic idea behind the algorithm is to schedule operations backwards in time. Initially, an operation is assigned to the instruction $-c(A)$, which is the latest time the operation can be scheduled for it to complete by time 0. Then, the nested loop repeatedly selects an operation A and one of its predecessors P and P's position $L(P)$ in the schedule is updated to be at least $c(P)$ cycles earlier than $L(A)$, the latest possible time at which

P can be scheduled to satisfy the dependence with A. Since operations are only moved earlier in the schedule to satisfy dependences, when no more operations need to be moved the schedule is a minimum length legal schedule in which every operation is scheduled as late as possible.

The final loop renumbers the instructions from the range $-m, \ldots, -1$ to the range $1, \ldots, m$; thus the total number of instructions in the schedule is m.

3.4 Limited Resources

In this section, we consider algorithms that deal with the reality that machines have limited functional resources. While scheduling the operations in the basic block $\mathcal{B}$, one now needs to worry about the potential resource conflicts between two operations in addition to any dependence constraints. While it is straightforward to give algorithms that compute optimal (shortest) schedules without resource constraints, finding shortest schedules with resource constraints is a much harder problem and, in practice, heuristics that give good but not necessarily the best schedules are used. Register allocation is not treated here. Ideally register allocation would be done simultaneously with scheduling as the registers are a special form of resource that can sometimes be renamed to improve scheduling, but current practice is to fix a register allocation either before or after scheduling. For simplicity, a pipelined implementation is assumed for each multi-cycle operation.

3.4.1 List Scheduling

List Scheduling employs a greedy approach to schedule as many operations as possible among those whose predecessors have already been scheduled. Each operation is assigned a priority. Operations ready for scheduling—meaning that all of the operations they depend on have already completed by the current control step of the schedule—are placed on a *ready list* ordered by their priorities. At each control step, the operation with the highest priority is scheduled first. If there are two or more operations with the same priority, then an arbitrary choice is made.

Algorithm 3.3. Given a basic block $\mathcal{B}$, its dependence graph, and a machine with limited resources, this algorithm finds a schedule for $\mathcal{B}$. For each $A \in \mathcal{B}$, the sets $\text{Pred}(A)$ and $\text{Succ}(A)$, as well as the labels $\ell(A)$ and $L(A)$ are assumed to have already been computed.

$V \leftarrow \mathcal{B}$
$k \leftarrow 1$
while $V \neq \emptyset$ **do**
  $\mathcal{S} \leftarrow$ all operations in V whose predecessors have finished executing before
      control step k
  for each $A \in \mathcal{S}$ **do**
    $\mu(A) \leftarrow L(A) - \ell(A)$
  endfor
  Sort $\mathcal{S}$ into a sequence $(A_{r_1}, A_{r_2}, \ldots, A_{r_{|S|}})$
    in increasing order of mobility $\mu(\cdot)$
  Create an empty instruction $\mathbf{I}_k$
  for $i = 1$ to $|S|$ **do**
    if A_{r_i} does not have resource conflicts with operations in $\mathbf{I}_j$, $1 \leq j \leq k$ **then**
      Put A_{r_i} in $\mathbf{I}_k$
      $V \leftarrow V - \{A_{r_i}\}$
      $\mathcal{S} \leftarrow \mathcal{S} - \{A_{r_i}\}$
    endif
  endfor
  $k \leftarrow k + 1$
  for each $A \in S$ **do**
    $\ell(A) \leftarrow \min(\ell(A) + 1, L(A))$
  endfor
endwhile

Figure 3.3 : List Scheduling Algorithm

List scheduling encompasses a family of different algorithms based on the choice of the priority function. In the algorithm described here (Figure 3.3), the priority of an operation A is defined by the difference $\mu(A) = L(A) - \ell(A)$ between its ALAP and ASAP labels, called the *mobility* of the operation. An operation with lower mobility has higher priority.

First, Algorithm 3.1 and Algorithm 3.2 are used to find the ASAP and ALAP labels of each operation in the given basic block. Operations without predecessors are placed on a ready list $\mathcal{S}$ and arranged in order of increasing mobility. They are taken from the ready list and scheduled one by one subject to the availability of machine resources. After one round, if there is still an operation A left over in $\mathcal{S}$, then its ASAP label is increased by 1 without exceeding its ALAP label. This reduces the mobility of A, if it is not already zero.

3.4.2 Linear Analysis

The Linear Algorithm (Figure 3.4) checks the operations of the basic block *linearly* in their order of appearance, and puts them into instructions observing dependence constraints and avoiding resource conflicts.

Algorithm 3.4. Given a basic block $(A_1, A_2, \ldots, A_n)$ of n operations, its dependence graph, and a machine with limited resources, this algorithm finds a schedule for the block. At any point in the algorithm, all scheduled operations lie in the sequence of instructions $(\mathbf{I}_t, \mathbf{I}_{t+1}, \ldots, \mathbf{I}_b)$, where $t \leq 1$ and $b \geq 1$.

```
t ← 1
b ← 1
Put the operation A_1 in instruction I_1
for i = 2 to n do
  k ← earliest control step ≥ t before which all predecessors of A_i finish executing
  while k ≤ b and A_i has a resource conflict with operations in I_k do
    k ← k + 1
  endwhile
  if k ≤ b then
    put A_i in I_k
  else
    if Pred(A_i) = ∅ then
      Put A_i in I_{t-1}
      t ← t − 1
    else
      Put A_i in I_k
      b ← k
    endif
  endif
endfor
```

Figure 3.4 : The Linear Algorithm

The operations in the block are first arranged in their prescribed sequential order: $A_1, A_2, \ldots, A_n$. To simplify the description of the algorithm, it is convenient to assume that a control step can be any integer. Start with an infinite sequence of empty instructions $\{\mathbf{I}_k : -\infty < k < \infty\}$ arranged in an imaginary vertical column. The operations $A_1, A_2, \ldots, A_n$ are taken in this order and put in instructions one by one. At any point, all operations already scheduled are in a range of instructions $(\mathbf{I}_t, \mathbf{I}_{t+1}, \ldots, \mathbf{I}_b)$, where $t \leq 1$ and $b \geq 1$. The value of t keeps decreasing and the value of b keeps increasing.

As with the ALAP algorithm (recall Figure 3.2), the final sequence of instructions $(\mathbf{I}_t, \mathbf{I}_{t+1}, \ldots, \mathbf{I}_b)$ can be renumbered to get a sequence of instructions $(\mathbf{I}'_1, \mathbf{I}'_2, \ldots, \mathbf{I}'_m)$, where $m = b - t + 1$.

Both t and b are initialized to 1. The first operation A_1 is put in instruction $\mathbf{I}_1$. Suppose operations $A_1, A_2, \ldots, A_{i-1}$ have already been scheduled, and the time has come to schedule A_i, where $2 \leq i \leq n$. Let k denote the smallest integer $\geq t$, such that all predecessors of A_i finish executing before control step k. If A_i has no predecessors, then $k = t$. Otherwise, if a predecessor A_r is in an instruction $\mathbf{I}_j$, where $t \leq j \leq b$, then $k \geq j + c(A_r)$. The exact value of k is found by taking the maximum of all such expressions.

So, operation A_i can be put in instruction $\mathbf{I}_k$ without violating any dependence constraints. If $k \leq b$, start checking the instructions $\mathbf{I}_k, \mathbf{I}_{k+1}, \ldots, \mathbf{I}_b$, in this order, for potential resource conflicts. If there is an instruction in this range with which A_i does not conflict, then put A_i in the first such instruction. Otherwise, A_i conflicts with all instructions in the range and $k = b + 1$. If $k > b$ was true before the checking could start, then its value did not change. At this point, if A_i had no predecessors in the first place, then put it at the top in a new instruction $\mathbf{I}_{t-1}$ and decrease t to $t - 1$. If A_i had predecessors, then put it in instruction $\mathbf{I}_k$ and increase b to k. The process ends when all operations in the given basic block have been scheduled.

3.5 An Example

Consider a basic block $\mathcal{B}$ consisting of 8 operations $A_1, A_2, \ldots, A_8$ arranged in this order. The dependence graph of $\mathcal{B}$ is given in Figure 3.5. The type, cycle time, the immediate predecessors, and the immediate successors of each operation are listed in Table 3.1. It is assumed that the cycle time of an addition is 1 and of a multiplication is 4. The four basic block scheduling algorithms we have described are applied to $\mathcal{B}$ one-by-one. The list scheduling and linear algorithms are customized for a machine with one adder and one multiplier. Note that the critical path in the dependence graph is $A_2 \rightarrow A_4 \rightarrow A_6 \rightarrow A_8$. Since this path's weight is $T_0 = 10$, the total number of cycles taken by any schedule for $\mathcal{B}$ must be at least 10.

ASAP Algorithm. At the beginning, the ASAP label $\ell(A_i)$ of each operation A_i is initialized to 1. The ASAP algorithm permits us to update

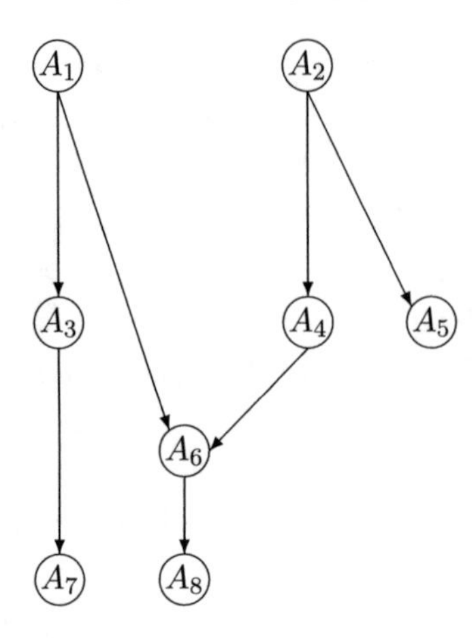

Figure 3.5 : Dependence Graph of Basic Block $\mathcal{B}$

OP A	Type	c(A)	Pred(A)	Succ(A)	$\ell(A)$	$L(A)$	$\mu(A)$
A_1	+	1		A_3, A_6	1	5	4
A_2	*	4		A_4, A_5	1	1	0
A_3	+	1	A_1	A_7	2	9	7
A_4	+	1	A_2	A_6	5	5	0
A_5	+	1	A_2		5	10	5
A_6	+	1	A_1, A_4	A_8	6	6	0
A_7	+	1	A_3		3	10	7
A_8	*	4	A_6		7	7	0

Table 3.1 : Details of the Basic Block of Figure 3.5

the labels of operations in any order and the order makes no difference to the final outcome. However, visiting the nodes in topological order is most efficient, as this order updates each operation's label exactly once for each of the operation's predecessors. The roots of the dependence graph are operations A_1 and A_2. These operations have no predecessors and so will retain their ASAP labels of 1 throughout the algorithm. We next consider operations that are nodes at depth 1 in the graph; these are operations whose longest chain of dependences from any root is a path of length 1. These operations A_3, A_4 and A_5 are exactly the ones that can be scheduled once the control steps of the roots A_1 and A_2 are known. The ASAP values of these operations are increased as follows:

$$\ell(A_3) \leftarrow \max\{\ell(A_3), \ell(A_1) + c(A_1)\} = \max\{1, 2\} = 2$$
$$\ell(A_4) \leftarrow \max\{\ell(A_4), \ell(A_2) + c(A_2)\} = \max\{1, 5\} = 5$$
$$\ell(A_5) \leftarrow \max\{\ell(A_5), \ell(A_2) + c(A_2)\} = \max\{1, 5\} = 5$$

We next consider operations at depth 2, those nodes in the graph whose longest chain of dependences from a root is a path of length 2. For the example, these operations are A_6 and A_7, whose ASAP values are increased as follows:

$$\ell(A_7) \leftarrow \max\{\ell(A_7), \ell(A_3) + c(A_3)\} = \max\{1, 3\} = 3$$
$$\ell(A_6) \leftarrow \max\{\ell(A_6), \ell(A_4) + c(A_4)\} = \max\{1, 6\} = 6.$$

Because A_6 has two predecessors we must also perform an update for its other predecessor A_1:

$$\ell(A_6) \leftarrow \max\{\ell(A_6), \ell(A_1) + c(A_1)\} = \max\{6, 2\} = 6.$$

The operation A_8 is the only operation at depth 3 in the dependence graph and is the last operation to be scheduled when the dependence graph is processed in topological order. Operation A_8's ASAP value is increased as follows:

$$\ell(A_8) \leftarrow \max\{\ell(A_8), \ell(A_6) + c(A_6)\} = \max\{1, 7\} = 7$$

It is easy to check at this point that $\ell(\cdot)$ satisfies all of the dependences and so is a valid schedule. Again, choosing a topological order to visit the operations guarantees the ASAP algorithm terminates in only a single pass over the nodes of the dependence graph, but visiting the nodes in any order will work, though possibly more slowly due to multiple visits to some nodes. All of the ASAP values are shown in Table 3.1.

The total number of control steps taken to execute $\mathcal{B}$ is

$$\max_{1 \le i \le 8} [\ell(A_i) + c(A_i)] - 1 = 10.$$

The total number of instructions in the schedule for $\mathcal{B}$ is $\max_i \ell(A_i) = 7$. The instructions are $(\mathbf{I}_1, \mathbf{I}_2, \ldots, \mathbf{I}_7)$, where $\mathbf{I}_4$ is empty, and

$$\mathbf{I}_1 = \{A_1, A_2\},\ \mathbf{I}_2 = \{A_3\},\ \mathbf{I}_3 = \{A_7\},\ \mathbf{I}_5 = \{A_4, A_5\},\ \mathbf{I}_6 = \{A_6\},\ \mathbf{I}_7 = \{A_8\}$$

The corresponding pseudo-assembly code is given below:

```
I1        A1, A2
I2        A3
I3        A7
I4        -
I5        A4, A5
I6        A6
I7        A8
```

ALAP Algorithm. At the beginning, $L(A_i)$ is initialized to $-c(A_i)$ for each A_i. As with the ASAP algorithm, we can visit the nodes in any order and obtain the correct answer, but some orders are more efficient than others. For the ALAP algorithm, processing the graph in a reverse topological order results in updating each operation just once for each of its successors.

We begin with the terminal nodes of the graph, operations A_5, A_7, and A_8 that have no successors. These nodes retain their initial ALAP labels of -1, -1, and -4, respectively. We next consider all operations that are reached by a longest backwards path of length 1; these are the operations A_3 and A_6 whose only successors are terminal nodes in the dependence graph. Their labels are updated as follows:

$$L(A_3) \leftarrow \min\{L(A_3), L(A_7) - c(A_3)\} = \min\{-1, -2\} = -2$$
$$L(A_6) \leftarrow \max\{L(A_6), L(A_8) - c(A_6)\} = \min\{-1, -5\} = -5$$

Next we consider the nodes whose longest backward path from a terminal node is 2, A_1 and A_4:

$$L(A_1) \leftarrow \min\{L(A_1), L(A_3) - c(A_1)\} = \min\{-1, -3\} = -3$$
$$L(A_4) \leftarrow \max\{L(A_4), L(A_6) - c(A_4)\} = \min\{-1, -6\} = -6$$

We must also perform an update for A_1's other successor, A_6:

$$L(A_1) \leftarrow \min\{L(A_1), L(A_6) - c(A_1)\} = \min\{-3, -6\} = -6$$

Finally, A_2 is the only node that is reachable from a terminal node via a backward path of length 3 and is therefore the last to be considered in any reverse topological order. Recall that, because $c(A_2) = 4$, operation A_2 is initially scheduled at time -4. There are two updates, one for each successor:

$$L(A_2) \leftarrow \min\{L(A_2), L(A_4) - c(A_2)\} = \min\{-4, -7\} = -10$$
$$L(A_2) \leftarrow \min\{L(A_2), L(A_5) - c(A_2)\} = \min\{-10, -5\} = -10$$

At this point the instructions are renumbered so that the first instruction occurs at control step 1. Since $m = -10$, we add $|m| + 1 = 11$ to the index of each instruction. Those instructions are $(\mathbf{I}_1, \mathbf{I}_2, \ldots, \mathbf{I}_{10})$, where $\mathbf{I}_2, \mathbf{I}_3, \mathbf{I}_4, \mathbf{I}_8$ are empty, and

$$\mathbf{I}_1 = \{A_2\},\ \mathbf{I}_5 = \{A_1, A_4\},\ \mathbf{I}_6 = \{A_6\},\ \mathbf{I}_7 = \{A_8\},\ \mathbf{I}_9 = \{A_3\},\ \mathbf{I}_{10} = \{A_5, A_7\}$$

The final values of $L(A)$ are given in Table 3.1. The pseudo-assembly code comparing ASAP and ALAP is given below:

```
           ASAP              ALAP
I1         A1, A2            A2
I2         A3                -
I3         A7                -
I4         -                 -
I5         A4, A5            A1, A4
I6         A6                A6
I7         A8                A8
I8         -                 -
I9         -                 A3
I10        -                 A5, A7
```

List Scheduling Algorithm. Initially, V has all 8 operations. The initial value of the mobility of each operation is listed in Table 3.1. At step 1, $\mathcal{S} = \{A_1, A_2\}$. The operations are arranged in the order (A_2, A_1), since $\mu(A_2) < \mu(A_1)$. Both operations can be put in instruction $\mathbf{I}_1$, since there is no resource conflict between them. At step 2, $\mathcal{S} = \{A_3\}$. So, A_3 goes into $\mathbf{I}_2$. Similarly, A_7 goes into $\mathbf{I}_3$. At step 4, $\mathcal{S} = \emptyset$. At step 5, $\mathcal{S} = \{A_4, A_5\}$. The operations are arranged in this order, since $\mu(A_4) < \mu(A_5)$. Only A_4 can be put into $\mathbf{I}_5$. We take A_4 out of $\mathcal{S}$ and decrease $\mu(A_5)$ to 4. At step 6, $\mathcal{S} = \{A_5, A_6\}$. The operations are arranged in the order (A_6, A_5), since $\mu(A_6) < \mu(A_5)$. Only A_6 can be put into $\mathbf{I}_6$ and $\mu(A_5)$ is decreased to 3. At step 7, $\mathcal{S} = \{A_5, A_8\}$. The operations are arranged in the order (A_8, A_5), since $\mu(A_8) < \mu(A_5)$. Both operations can be put in instruction $\mathbf{I}_7$, since there is no resource conflict between them. The total number of instructions in the schedule for $\mathcal{B}$ is 7. Those instructions are $(\mathbf{I}_1, \mathbf{I}_2, \ldots, \mathbf{I}_7)$, where $\mathbf{I}_4$ is empty, and

$$\mathbf{I}_1 = \{A_1, A_2\}, \mathbf{I}_2 = \{A_3\}, \mathbf{I}_3 = \{A_7\}, \mathbf{I}_5 = \{A_4\}, \mathbf{I}_6 = \{A_6\}, \mathbf{I}_7 = \{A_5, A_8\}$$

The total number of cycles needed is $[7+\max\{c(A_5), c(A_8)\}-1] = 10$.

Linear Algorithm. Initially, $t = b = 1$. Put operation A_1 in instruction $\mathbf{I}_1$. Since A_2 has no predecessors and does not conflict with A_1, put A_2 also in $\mathbf{I}_1$. Operation A_3 has only one predecessor, namely A_1. Since $c(A_1) = 1$, A_3 can be put in instruction $\mathbf{I}_2$. Increase b to 2. The sole predecessor of A_4 is A_2 and $c(A_2) = 4$. Hence, A_4 can go into instruction $\mathbf{I}_5$. Increase b to 5. Operation A_5 also has A_2 as its only predecessor. But, A_5 cannot go into $\mathbf{I}_5$, since it conflicts with A_4. So, put A_5 in $\mathbf{I}_6$, and increase b to 6. Operation A_6 has two predecessors: A_1 and A_4. The earliest instruction that can take A_6 without violating dependence constraints is $\mathbf{I}_6$. But $\mathbf{I}_6$ already has A_5 and it conflicts with A_6. So, put A_6 in $\mathbf{I}_7$ and increase b To 7. Now it is easy to see that A_7 can go into $\mathbf{I}_3$ and A_8 into $\mathbf{I}_8$. Increase b to 8.

The total number of instructions in the schedule for $\mathcal{B}$ is 8. Those instructions are $(\mathbf{I}_1, \mathbf{I}_2, \ldots, \mathbf{I}_8)$, where $\mathbf{I}_4$ is empty, and

$$\mathbf{I}_1 = \{A_1, A_2\}, \mathbf{I}_2 = \{A_3\}, \mathbf{I}_3 = \{A_7\}, \mathbf{I}_5 = \{A_4\}, \mathbf{I}_6 = \{A_5\}, \mathbf{I}_7 = \{A_6\}, \mathbf{I}_8 = \{A_8\}$$

The total number of cycles needed is $[8 + c(A_8) - 1] = 11$. It is worth pointing out that while the linear algorithm produces a schedule that is longer than list scheduling for this example, that does not mean that the linear algorithm is worse in general; there are also examples where the linear algorithm performs better than list scheduling.

3.6 More Algorithms

3.6.1 Critical Path Algorithm

The *weight* of a path in a dependence graph of operations is the sum of execution times of all operations on that path. A *critical path* is a path with the highest weight. A *critical operation* is an operation on a critical path.

Let T denote the minimum number of control steps needed to completely execute all operations in the given program, ignoring resource constraints (i.e., as computed by the the ASAP and ALAP algorithms). For each $A \in \mathcal{B}$, we have $1 \leq \ell(A) \leq L(A) \leq T$, where $\ell(A)$ denotes the ASAP label of A and $L(A)$ its ALAP label. An operation A is critical iff $\ell(A) = L(A)$. For $1 \leq i \leq T$, the ith *critical frame*, denoted by C_i, is the set of all critical operations A such that $\ell(A) = L(A) = i$. Some of these frames may be empty.

First, we apply the ASAP and ALAP algorithms to find the ASAP and ALAP labels of each operation in the given program. Then we find the critical frames by matching those labels. Although the operations in a given frame are independent of each other, there may not be enough available resources to execute them all at once. We apply a form of list scheduling to the operations of a frame C_i to create a sequence of instructions which we call the sub-frames of C_i and denote by $C_{i1}, C_{i2}, \ldots, C_{iT_i}$.

The first step is to choose a weighting function that can be used to linearly order a set of operations that are ready to be scheduled. This choice affects the speed at which our algorithm will run. Extensive tests of different functions are given in [Fis79]. It has been reported that excellent results are obtained with the function that assigns a weight to each operation A that is the length (in number of operations) of the longest chain in the dependence graph from A to a terminal node [LDSM80]. A very simple weighting function is the one that assigns to each operation its numerical position in the original sequential program.

The operations in C_i are arranged in decreasing order of their weights. The first operation (i.e., the one with the highest weight) is moved to the first sub-frame C_{i1} of C_i, which is initially empty. The second operation in C_i is then checked. If there is no resource conflict between it and the operations already in C_{i1}, it is also moved to C_{i1}. Otherwise, we check the third operation in C_i; and so on. When no more operations can be moved to C_{i1}, we have found all operations that can be put in the first sub-frame C_{i1}. These operations are deleted from C_i, and we look for operations in this reduced set that can be put in the next sub-frame C_{i2}, that is, can be scheduled in the next control step. The full algorithm is given in Figure 3.6. Once this algorithm completes, only the non-critical operations remain to be scheduled using any of the techniques discussed previously.

```
Algorithm 3.5.
Find the ASAP label ℓ(v) for each v ∈ V
Find the ALAP label L(v) for each v ∈ V, and the minimum number T
of steps needed to execute the program
for i = 1 to T do
   C_i ← ∅
endfor
for each v ∈ V do
   // Allocate critical operations to the critical frames
   if ℓ(v) = L(v) then
      put v in C_ℓ(v)
   endif
endfor
// Schedule critical frames
for i = 1 to T do
   arrange the operations in C_i in decreasing order of weight
   j ← 0
   // Schedule critical operations
   while C_i ≠ ∅ do
      j ← j + 1
      C_ij ← ∅
      k ← 0
      while (operations in C_i remain to be examined) do
         k ← k + 1
         if (there is no resource conflict between the k^th operation v of C_i
            and the operations in C_ij) then
            add v to C_ij
         endif
      endwhile
      remove the operations of C_ij from C_i
   endwhile
   T_i ← j
endfor
```

Figure 3.6 : Critical path scheduling algorithm for limited resources

---

Example 3.1. Consider the data flow graph, $G(V, E)$ in Figure 3.7(a). We assume one adder and one multiplier are available in the resource library for scheduling G in Figure 3.7(a). First, we determine the critical frames of G by finding its ASAP schedule - Figure 3.7(b), and its ALAP schedule - Figure 3.7(c), assuming unlimited resources.

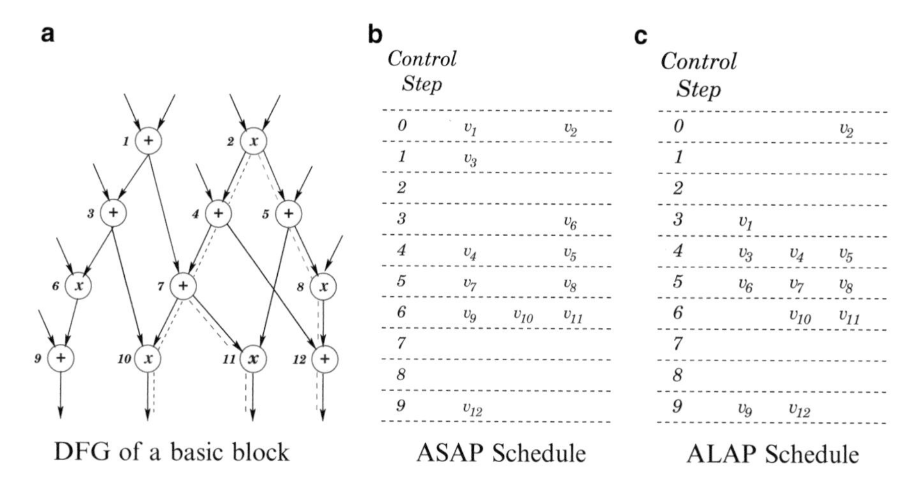

Figure 3.7 : A graph G and its ASAP, ALAP schedules

Critical Frame	Critical Operations
C_0	v_2
C_4	v_4, v_5
C_5	v_7, v_8
C_6	v_{10}, v_{11}
C_9	v_{12}

Table 3.2 : Critical frames of G

The critical path(s) of G are shown with dashed lines in Figure 3.7(a). Table 3.2 lists all the critical frames of G and their constituent critical operations. Though operations v_4 and v_5 in critical frame C_4 are independent of each other, they cannot be scheduled in parallel as only one adder is available. We create subframes of C_4, namely C_{4_1} and C_{4_2}, containing operations v_4 and v_5 respectively.

The final resource-constrained schedule of G (Figure 3.7(a)) is shown in Figure 3.8.

Control Step	◯	ⓧ
0	v_1	v_2
1	v_3	
2		
3		
4	v_4	v_6
5	v_5	
6	v_7	
7		
8	v_9	v_8
9		
10		
11		
12	v_{12}	v_{10}
13		
14		
15		
16		v_{11}

Figure 3.8 : Resource-constrained schedule of DFG in Figure 3.7(a)

3.6.2 Restricted Branch and Bound Algorithm

"Branch and bound" is the name given to a class of search tree algorithms where the goal is to find one best path through the search tree according to some metric. Instruction scheduling can be thought of as a search tree problem in the following way: we have a choice of the first instruction to schedule, and given that decision we then have a choice of the second instruction to schedule, and so on. Clearly, this structure defines a search tree, with each complete path representing an order on the scheduling of all instructions, and furthermore some path (or paths) will be a minimum cost schedule.

Because of the exponential number of possible orderings of instructions in general, people have sought methods to avoid exhaustively exploring all paths of the search tree. (As an aside, the "bound" part of branch and bound is one such method, which cuts off the exploration of any path once its cost exceeds that of the lowest cost path found so far.) In instruction scheduling, a standard approach has been to assign weights to instructions, and to prioritize the scheduling of high weight instructions above those with lower weight. In the case where all instructions have distinct weights, this reduces to a list scheduling algorithm—there are no choices in the ordering of instructions and so only one branch of the search tree is explored, and there is no branching. Conversely, when all instructions have the same weight, we have a pure search tree problem in which any order that respects instruction dependences can be considered. When some weights are distinct and some are the same we have a restricted search problem in which only some paths of the full search tree are considered (thus the name, "restricted branch and bound").

For simplicity, the algorithm we consider here assumes a weighting function has been chosen that linearly orders the operations that are ready to be scheduled—i.e., all the weights are distinct. Thus, we will describe a list scheduling algorithm, but it is straightforward to generalize this to branch-and-bound on a search tree as outlined above. In a given cycle i, we find the set $\mathcal{S}_i$ of all unscheduled operations that are mutually independent and have no unfinished immediate predecessors. The operations in $\mathcal{S}_i$ are then arranged in decreasing order of their weights. The first operation (i.e., the one with the highest weight) is moved to the ith instruction $\mathcal{I}_i$ which is initially empty. The second operation in $\mathcal{S}_i$ is then checked. If there is no resource conflict between it and the operations already in $\mathcal{I}_i$, it is also moved to $\mathcal{I}_i$. Otherwise, we check the third operation in $\mathcal{S}_i$; and so on. When no more operations can be moved to $\mathcal{I}_i$, we have found all operations that can be scheduled in control step i. These operations are deleted from V. The remaining operations of set $\mathcal{S}_i$ are then merged with $\mathcal{S}_{i+1}$. The updated set $\mathcal{S}_{i+1}$ is then scheduled in the next control step.

The input to this algorithm is the dependence graph $G(V, E)$ of the basic block, the function c giving the cycle times of the operations (we will present this algorithm for operations that can take more than one cycle) and a list of necessary resources for each operation. The output consists of a sequence of instructions $\mathcal{I}_1, \mathcal{I}_2, \ldots, \mathcal{I}_N$ such that

- Either an instruction is empty or it contains some operations from V. Each operation in V belongs to exactly one of the N instructions.

- In cycle i, sufficient resources are available to simultaneously execute all operations in $\mathcal{I}_i$.

Algorithm 3.6.
// Determine predecessor count and successor set
for each $v \in V$ **do**
  compute $\mathrm{Pred}(v)$ and $\mathrm{Succ}(v)$
  $\mathrm{WAIT}(v) \leftarrow 1$
endfor
$i \leftarrow 0$
while $V \neq \emptyset$ **do**
  $i \leftarrow i + 1$
  for each $v \in V$ **do**
    $\mathrm{WAIT}(v) \leftarrow \mathrm{WAIT}(v) - 1$
  endfor
  // Determine ready operations
  $\mathcal{S}_i \leftarrow \{v \in V : \mathrm{Pred}(v) = \emptyset \text{ and } \mathrm{WAIT}(v) \leq 0\}$
  arrange the operations in $\mathcal{S}_i$ in decreasing order of weight
  $j \leftarrow 0$
  $\mathcal{I}_i \leftarrow \emptyset$
  while (operations in $\mathcal{S}_i$ remain to be examined) **do**
    $j \leftarrow j + 1$
    // Check for resource conflict
    if (enough resources are available at control step i to execute
      the operations in $\mathcal{I}_i$ along with the jth operation of $\mathcal{S}_i$) **then**
        put that operation in $\mathcal{I}_i$
    endif
  endwhile
  for each $v \in \mathcal{I}_i$ **do**
    for each $w \in \mathrm{Succ}(v)$ **do**
      $\mathrm{Pred}(w) \leftarrow \mathrm{Pred}(w) - v$
      $\mathrm{WAIT}(w) \leftarrow \max(\mathrm{WAIT}(w), c(v))$
    endfor
  endfor
  $V \leftarrow V - \mathcal{I}_i$
endwhile
$N \leftarrow i$

Figure 3.9 : Restricted Branch and Bound Scheduling (RBBS) algorithm for limited resources

- An operation in the first instruction $\mathcal{I}_1$ does not depend on any other operation in V. All operations in $\mathcal{I}_1$ can start executing as soon as we are ready to execute the given basic block.

- For $2 \leq i \leq N$, an operation in $\mathcal{I}_i$ may depend on operations in the instructions $\mathcal{I}_1, \mathcal{I}_2, \ldots, \mathcal{I}_{i-1}$, but not on any operation in the instructions $\mathcal{I}_i, \mathcal{I}_{i+1}, \ldots, \mathcal{I}_N$. All operations in $\mathcal{I}_i$ can start executing in parallel in the ith cycle after the start of the basic block.

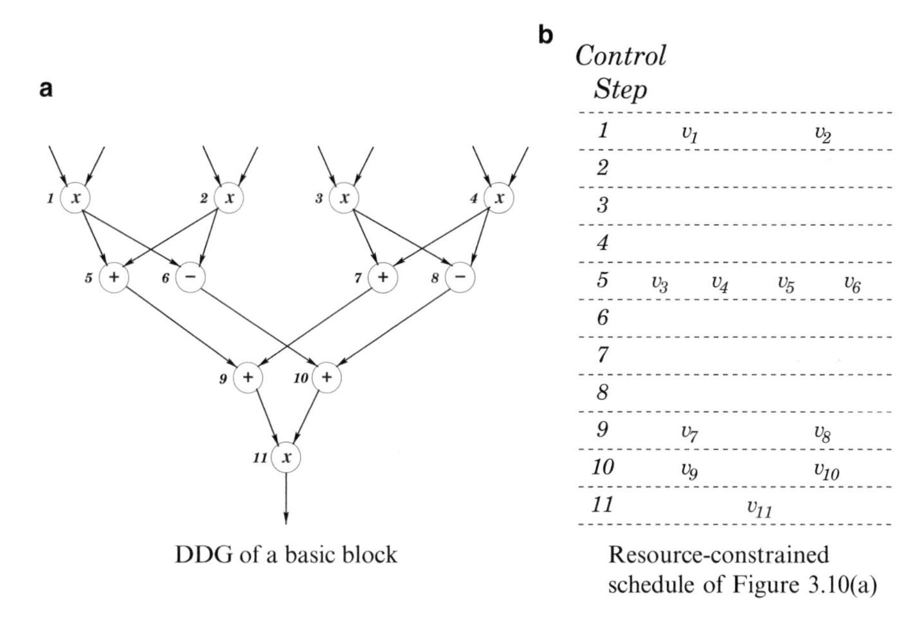

Figure 3.10 : Restricted Branch and Bound Scheduling

Example 3.2. Consider the data dependence graph (DDG), $G(V, E)$ in Figure 3.10(a). We assume two adders (each capable of performing addition or subtraction) and two multipliers are available in the resource library for scheduling G in Figure 3.10(a). First, we compute the sets $\mathcal{S}_0 \ldots \mathcal{S}_{10}$ as defined in Figure 3.9. The variable WAIT restricts the inclusion of an operation v_i unless all of its immediate predecessors have finished execution. Table 3.3 lists all such sets for the data flow graph in Figure 3.10(a).

The operations in $\mathcal{S}_i$ are then arranged in decreasing order of weight. Each operation in $\mathcal{S}_i$ is assigned to instruction $\mathcal{I}_i$ if and only if enough resources are available. From Table 3.3, we observe that only two operations from the set S_0 can be put in $\mathcal{I}_0$ as there are only two multipliers available. Operations in $\mathcal{S}_i$ which cannot be put in $\mathcal{I}_i$ are merged with $\mathcal{S}_{i+1}$. Thus, $\mathcal{S}_i$ is dynamically updated based on the status of $\mathcal{S}_{i-1}$.

	Operations
S_1	v_1 v_2 v_3 v_4
S_2	v_3 v_4
S_3	v_3 v_4
S_4	v_3 v_4
S_5	v_3 v_4 $v_5 v_6$
S_6	
S_7	
S_8	
S_9	v_7 v_8
S_{10}	v_9 v_{10}
S_{11}	v_{11}

Table 3.3 : Sets S_i

Note that operation selection, in case two operations have the same weight, during instruction formation in RBBS affects the quality of the schedule. For instance, if operations v_1 and v_3 are put in $\mathcal{I}_1$ and operations v_2 and v_4 are put in $\mathcal{I}_2$, then the issue of v_5 and v_6 will be delayed until control step 9. Alternatively, a parallel issue of v_1 and v_2, as shown in Figure 3.10(a), allows the issue of v_5 and v_6 in control step 5.

3.6.3 Force-Directed Scheduling

The ASAP and ALAP algorithms schedule operations such that the total number of steps needed to complete execution of the whole program is minimized. However, such greedy approaches can lead to schedules where some instructions use many functional units and other instructions use very few. In situations where there are no fixed resource constraints it may be desirable to minimize a schedule's *load*, which is the maximum number of functional units required. For example, in designing hardware the problem of how to realize a basic block as a hardware circuit is essentially the basic block scheduling problem and minimizing the load of the selected schedule improves the utilization of the functional units compared to a pure greedy schedule. Force-directed scheduling [PK89] attempts to reduce the number of functional units required without lengthening execution time by spreading operations over the instructions so that every instruction uses as many of the same functional units as possible.

In a schedule that executes in minimum time, each operation $v \in V$ is scheduled in some control step between its ASAP label $\ell(v)$ and its ALAP label $L(v)$. Let the *slack* of v be $L(v) - \ell(v)$. Initially, force-directed scheduling assumes that v could end up being scheduled in any control step in the range $[\ell(v)..L(v)]$. The algorithm models this uncertainty by setting $p(v, i)$, the probability that v is scheduled in control step i, as follows:

$$p(v,i) = \begin{cases} 1/(\mathit{slack}(v)+1) & \text{if } \ell(v) \le i \le L(v) \\ 0 & \text{otherwise} \end{cases} \tag{3.1}$$

Note that if the slack is 0 (i.e., v is critical, recall subsection 3.6.1) then v must be scheduled in control step $L(v)$ (equivalently $\ell(v)$) and $p(v, L(v)) = 1$ and $p(v, j) = 0$ for every $j \neq L(v)$. For non-critical operations, the non-zero probability assigned to any control step decreases as the slack increases, reflecting that there are more possible control steps where the operation potentially could be scheduled.

Example 3.3. We illustrate the assignment of probabilities $p(v, i)$ for the data dependence graph in Figure 3.11(a). While we will only present the force-scheduling algorithm for single cycle operations, this example uses multicycle operations and the needed extension is not difficult. In Figure 3.11(b), the width of a box v in control step i reflects the probability of scheduling some stage of an operation v at

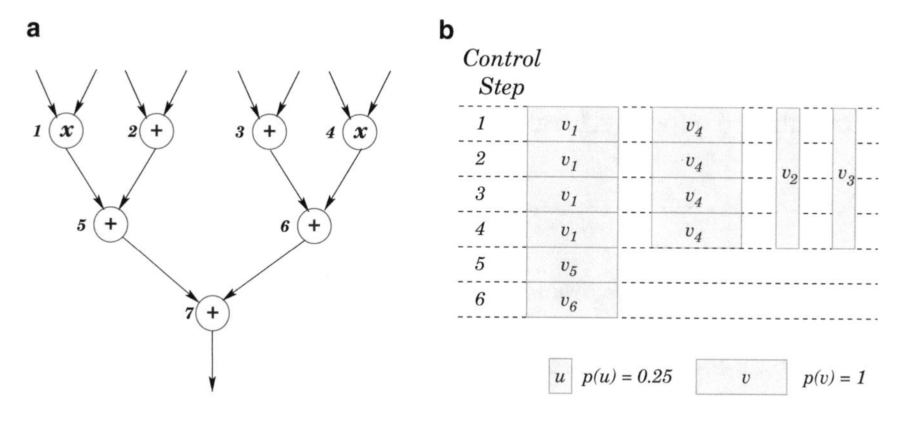

Figure 3.11 : Scheduling probability distribution

control step i. Four-cycle multiply operation v_1 with $\ell(v_1) = L(v_1) = 1$ is given probability 1 for control steps $1, 2, 3, 4$—this operation must start in step 1 and will run until step 4. Operation v_4 is similar. Single cycle add operations v_5, v_6 are given probability 1 for control steps $5, 6$. Operations v_2, v_3 are assigned probabilities 0.25 for the control steps $1, 2, 3, 4$ as they can be scheduled in any of these control steps without violating any data dependences.

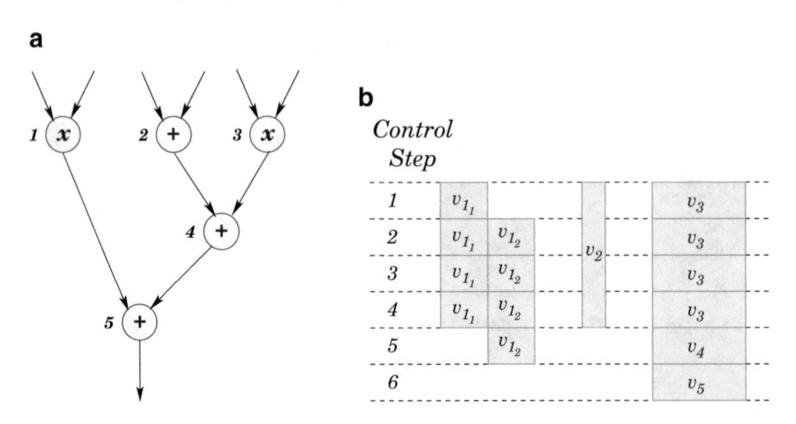

Figure 3.12 : Scheduling probability of non-critical multi-cycle operations

Now consider the data dependence graph shown in Figure 3.12(a). Here, the multiply operation v_1 can be scheduled either in control step 1 or 2. Thus, the probability of scheduling the first stage of v_1 in either control step is 0.5. However, if the first stage of v_1 is scheduled in control step 1 then the second cycle of v_1 executes in control step 2; in either case it is guaranteed that a stage of v_1 is executing in the second control step. Applying similar reasoning to the rest of the stages, we see that the combined probability distribution of all the stages is $0.5, 1.0, 1.0, 1.0, 0.5$. In Figure 3.12(a), the two possible choices of scheduling the four cycle operation v_1 are labeled v_{1_1} and v_{1_2}.

The key idea in force-directed scheduling is to use the *weight* of each control step to help guide whether to schedule operations in that control step or elsewhere. Let the type *typ*(v) of operation v name what kind of resource it uses (e.g., multiplier, adder—for simplicity we assume that operations use only a single resource). Then the weight

$W(i,t)$ of operations of type t in control step i is

$$\mathrm{W}(i,t) = \sum_{\{v \in V | typ(v)=t\}} p(v,i)$$

Intuitively, the weight of a control step represents the average number of operations in that control step over all possible legal schedules that meet the time bound, assuming that every operation can be scheduled anywhere between its ASAP and ALAP labels independently of any other operation. Of course, that assumption is potentially a gross approximation, but an approximation that is still useful as it gives an inexpensive measure of how many operations may end up in a control step if scheduling is done without considering instruction load.

Force-directed scheduling iteratively schedules one operation at a time. After computing the weight of each control step, the algorithm considers the effect of scheduling each operation $v \in V$ in every control step between $\ell(v)$ and $L(v)$. For each operation v and choice of control step i, the algorithm calculates whether placing v in i will have a positive or negative effect on the overall load when scheduling is complete. Scheduling v in instruction i concentrates all the probability for v in step i and reduces the probability of v being scheduled elsewhere to 0. Thus, scheduling v in step i will increase the load in step i and decrease it in other control steps. Force-directed scheduling calculates a *force* to model the change in potential load at each control step; the sum over all control steps is the total force of the decision to schedule operation v in step i. The final choice of which operation to schedule in which control step is done by choosing the option that minimizes the total force, which has the effect of avoiding placing operations in control steps that are likely to have many other operations that need to be scheduled there unless absolutely necessary. The entire process is repeated until every operation is scheduled.

To give good results, it turns out that it is important to model not just the change in potential load caused by concentrating all the probability for an operation in a single control step, but also to account for the effect this decision has on the immediate dependence predecessors and successors of the operation. Once an operation that previously had non-zero slack is scheduled in a specific control step, the range of possible control steps where its predecessors and successors can be scheduled may shrink, leading to concentration of their probability mass as well in fewer control steps. Thus, the full algorithm also calculates a total force that includes contributions from dependence predecessors and successors of the operation being scheduled.

We now define the rest of the terms used in Algorithm 3.1. **Self-force** represents the effect of scheduling an operation v at a control step i on the overall schedule:

$$\mathcal{F}_{\text{Self}}(v,i) = \sum_{\ell(v) \leq j \leq L(v)} W(j, typ(v))x(j) \tag{3.2}$$

where x is the increase (or decrease) in the probability of scheduling v at control step j due to scheduling v at control step i. For example, if the probability of scheduling v at control step i or $i+1$ was .5 each, then $x(i) = .5$ and $x(i+1) = -.5$.

Predecessor force represents the effect of scheduling an operation v at a control step i on its predecessors:

$$\mathcal{F}_{\text{Pred}}(v,i) = \sum_{v'} \sum_{j} W(j, typ(v))x' \quad \text{where} \quad v' \in \text{Pred}(v) \tag{3.3}$$

where j is any control step in which the predecessor v' may be scheduled given that v is scheduled in step i. Here x' is the increase (decrease) in the probability of scheduling the predecessor v' at control step j due to scheduling v at control step i.

Successor force represents the effect of scheduling an operation v at a control step i on its successors:

$$\mathcal{F}_{\text{Succ}}(v,i) = \sum_{v''} \sum_{k} W(k, typ(v))x'' \quad \text{where} \quad v'' \in \text{Succ}(v) \tag{3.4}$$

where k is any control step in which the successor v'' may be scheduled given that v is scheduled in step i. Here x'' is the increase (decrease) in the probability of scheduling a successor x'' at control step k due to scheduling v at control step i.

Total force represents the overall effect of scheduling an operation v on the resulting schedule. A negative value reflects improved distribution of concurrency (i.e., reduced load) in the resulting schedule. It is calculated as follows :

$$\mathcal{F}_{rmTotal}(v,i) = \mathcal{F}_{\text{Self}}(v,i) + \mathcal{F}_{\text{Pred}}(v,i) + \mathcal{F}_{\text{Succ}}(v,i) \tag{3.5}$$

Algorithm 3.1. Force-Directed Scheduling

while $V \neq \emptyset$ **do**
  $F_{min} \leftarrow \infty$
  $s \leftarrow \infty$
  $vv \leftarrow nil$
  // Compute ASAP, ALAP labels of each operation
  for each $v \in V$ **do**
    compute $\ell(v)$, $L(v)$
  endfor
  for each control step i **do**
    for each type of operation t **do**
      compute $\mathrm{W}(i, t)$
    endfor
  endfor
  for each $v \in V$ **do**
    for each control step i in $[\ell(v), L(v)]$ **do**
      compute $\mathcal{F}_{Total}(v, i)$ using Equation 3.5
      if $F_{min} > \mathcal{F}_{Total}(v, i)$ **then**
        $F_{min} = \mathcal{F}_{Total}(v)$
        $s \leftarrow i$
        $vv \leftarrow v$
      endif
    endfor
  endfor
  Schedule vv at control step s
  $V \leftarrow V - vv$
endwhile

Note that after each operation is scheduled the probability distributions for each remaining operation and the weight of each control step are recomputed.

Example 3.4. Consider the example in Figure 3.7(a). The ASAP and ALAP schedules are shown in Figure 3.7(b) and Figure 3.7(c) respectively. Figure 3.13(a) shows the initial probability distribution of all the operations, determined from their ASAP and ALAP schedules. Figure 3.13(b) shows the execution probabilities of operation 6.

Let us consider scheduling of v_1 in control step 2. The probability of execution of v_1 changes from $1/4$ to 1 for control step 2 and from $1/4$ to 0 for control steps $2, 3$ and 4. Thus, the self force is given by:

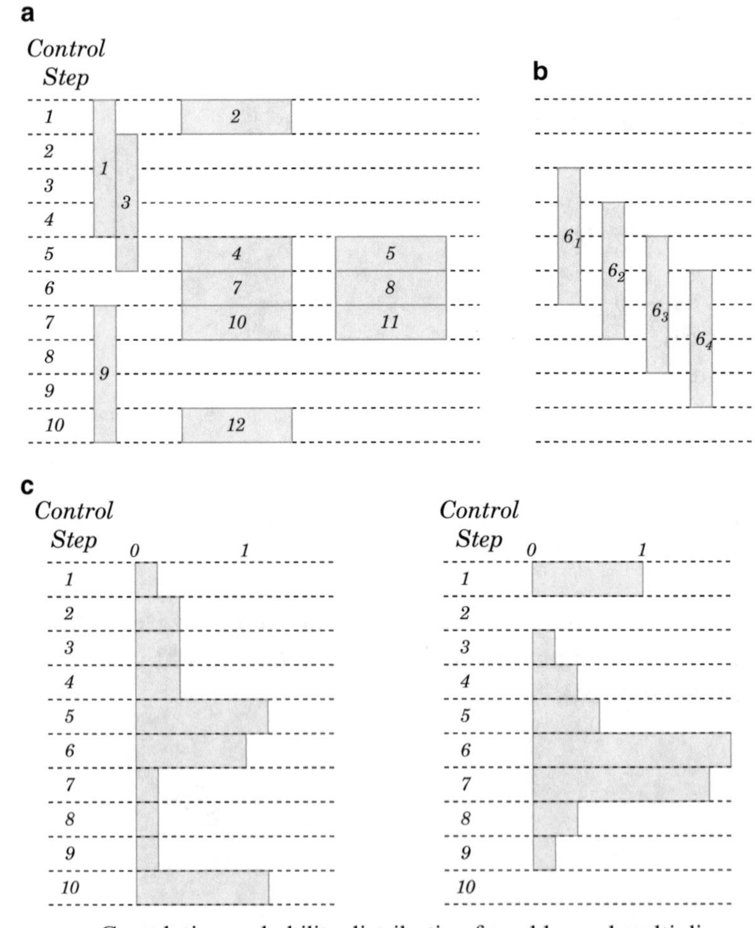

Figure 3.13 : Initial state

$$\begin{aligned}
\mathcal{F}_{\text{Self}} &= \mathcal{F}_{\text{Self}_1} + \mathcal{F}_{\text{Self}_2} + \mathcal{F}_{\text{Self}_3} + \mathcal{F}_{\text{Self}_4} \\
&= 0.25 \times (0.75) + 0.5 \times (-0.25) + 0.5 \times (-0.25) + 0.5 \times (-0.25) \\
&= -0.1875 \\
\mathcal{F}_{\text{Pred}} &= 0 \\
\mathcal{F}_{\text{Succ}} &= 0 \\
\mathcal{F}_{\text{Total}} &= \mathcal{F}_{\text{Self}} + \mathcal{F}_{\text{Pred}} + \mathcal{F}_{\text{Succ}} \\
&= -0.1875
\end{aligned}$$

3.7 Limited Beyond Basic Block Optimization

In the previous sections we discussed various approaches to parallelize basic blocks, both with and without resource constraints. However, in general, the amount of ILP within basic blocks is very limited. Thus, it is natural to look beyond basic blocks for further parallelization [Fis79, Fis81b, BR91, FGL94]. In this section we illustrate the beyond-basic-block optimizations described by Tokoro et al. [TTT81]. More aggressive techniques for scheduling beyond basic blocks are presented in the next chapter.

Unlike basic block optimizations, beyond basic block optimizations move an operation or operations across basic blocks, subject to data dependences [Ban76] and resource constraints. Let B denote the set of basic blocks in the control flow graph of the program. Let an operation be transferred from basic block b_i to basic block b_j, where $b_i, b_j \in B$. We refer the basic block b_i as the *source*, denoted by S_s and b_j as the *destination*, denoted by S_d. A destination can either be an immediate predecessor or an immediate successor basic block of the source; a source may have multiple destinations, denoted by S_{d_i} or a destination may have multiple sources, denoted by S_{s_i}. Next, we discuss different types of beyond basic block optimizations, i.e., transfer of operations between S_{s_i} and S_{d_i}.

Simple type: Consider the example shown in Figure 3.14, where S_s and S_d are adjacent basic blocks. Assume there is no data dependence between operations v_A, v_B. A *simple* transfer, denoted by $\mathcal{G}$(S), moves

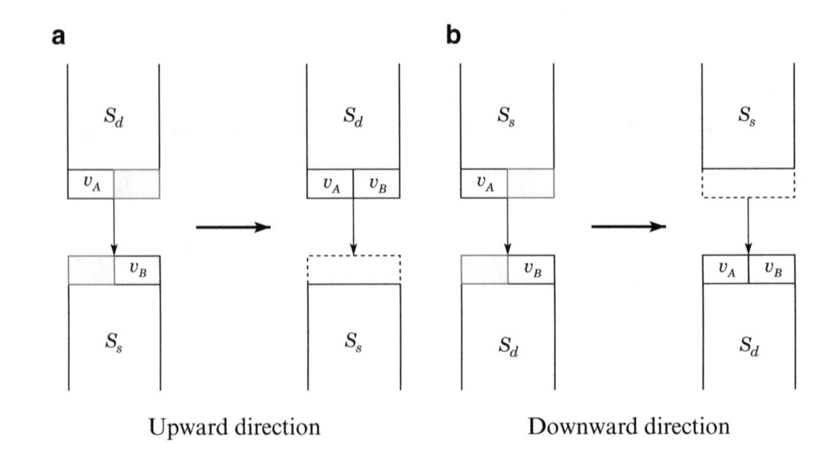

Figure 3.14 : Simple type transfer

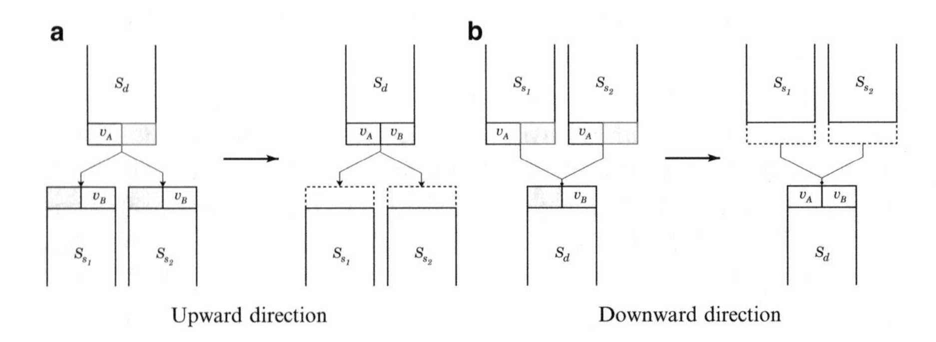

Figure 3.15 : Common type transfer

an operation either *upwards*, e.g., v_B in Figure 3.14(a) or *downwards*, e.g., v_A in Figure 3.14(b). $\mathcal{G}$(S) is legal *iff* Pred(v_B) in Figure 3.14(a) (Succ(v_A) in Figure 3.14(b)) does not contain any operation lying between v_A and v_B.

Common type: A *common* type transfer, denoted by $\mathcal{G}$(C), moves an operation v to S_d from multiple sources. As an illustration, consider the example shown in Figure 3.15. In Figure 3.15(a), $\mathcal{G}$(C) moves the common operation v_B *upwards*, whereas it moves v_A *downwards* in Figure 3.15(b). $\mathcal{G}$(C) is legal *iff* Pred(v_B) in Figure 3.15(a) (Succ(v_A) in Figure 3.15(b)) does not contain any operation lying between v_A and v_B.

Redundancy Reduction type: Eliminating a redundant operation from a source is denoted by $\mathcal{G}$(RR). An operation v is *redundant* if

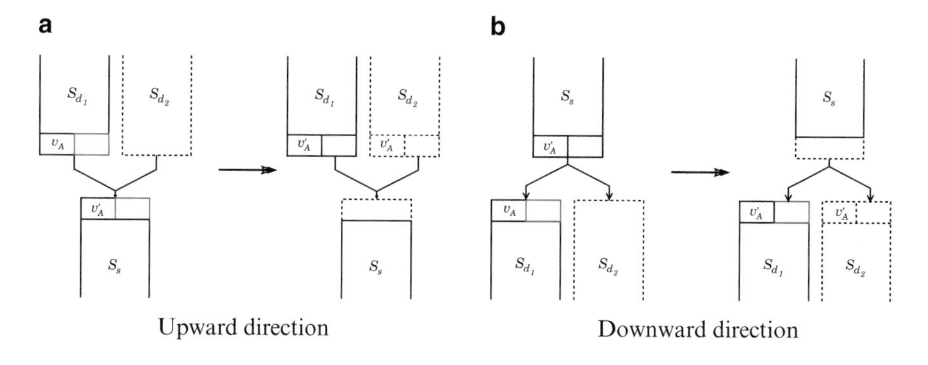

Figure 3.16 : Redundancy Reduction type transfer

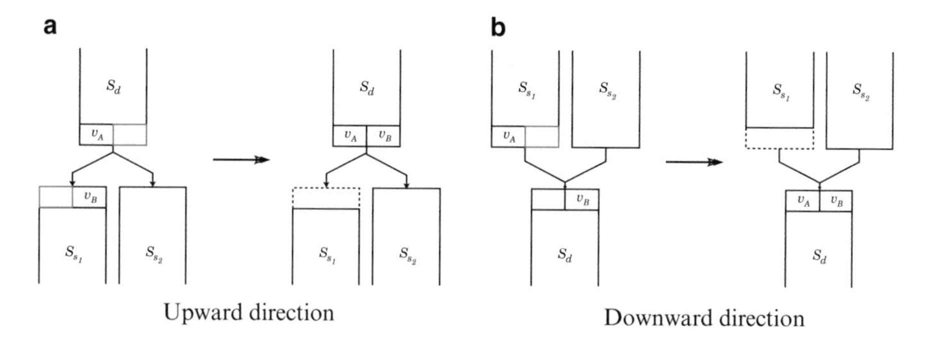

Figure 3.17 : Redundancy Insertion type transfer

the register or memory location written to by v is dead (is never referenced subsequently). As an illustration, consider the example shown in Figure 3.16. In Figure 3.16(a), $\mathcal{G}(\mathrm{RR})$ eliminates the redundant operation v_A and moves the operation $v_{A'}$ *upwards*. Similarly, in Figure 3.16(a), it eliminates the redundant operation v_A and moves the operation $v_{A'}$ *downwards*. $\mathcal{G}(\mathrm{RR})$ is legal *iff* $\mathrm{Pred}(v_{A'})$ in Figure 3.16(a) ($\mathrm{Succ}(v_{A'})$ in Figure 3.16(b)) does not contain any operation lying between v_A and $v_{A'}$.

Redundancy insertion type: Inserting a redundant operation in a destination basic block is denoted by $\mathcal{G}(\mathrm{RI})$. Consider the example shown in Figure 3.17. In Figure 3.17(a), $\mathcal{G}(\mathrm{RI})$ moves v_B *upwards*, from S_{s_1} to S_d. After this transformation, the operation v_B is now executed on the path to basic block S_{s_2} whereas prior to the transformation it was never executed on this particular path. Similarly, $\mathcal{G}(\mathrm{RI})$ moves v_A *downwards*, in Figure 3.17(b). $\mathcal{G}(\mathrm{RI})$ is legal *iff*

i) The contents written by v_B are read only by operations in S_{s_1}.

ii) $\mathrm{Pred}(v_B)$ in Figure 3.17(a) ($\mathrm{Succ}(v_A)$ in Figure 3.17(b)) does not contain any operation lying between v_A and v_B.

The optimizations described above (cursorily described for historical purposes) are subsumed by the techniques discussed in the subsequent chapters.

FURTHER READING

Basic block parallelization was first studied in the context of microprogramming. The book by Agerwala and Rauscher [AR76] gives a good introduction to microprogramming. The theory of job scheduling in operations research turned out to be a rich source of algorithms for the microprogramming research community [Hu61, CMM67, Hus70, CG72, RCG72, Bak74, Cof76, Gon77]). For linear analysis and list scheduling as covered in this book, a good place to start is the paper by Landskov et al. [LDSM80]. (The algorithms presented above try to factor in explicitly the cycle times of operations.) Other standard references are [Age76, DT76, Ban11a].

One of the early works in detection and parallel execution of independent instructions is due to Tjaden and Flynn [TF70]. The impact of conditionals on exploitation of parallelism was arguably first discussed by Riseman and Foster [RF72]; percolation of code for boosting parallel dispatching and execution was first discussed by Foster and Riseman in [FR72]. Discussion on identification of parallelism in microprograms is available in [JD74, DT76, Das77, Bar78, Har87]. Approaches for microprogram optimization and packing into micro-instruction words is discussed in [TTTT77, Woo78]. The late 1970s saw the development of techniques for global optimization of microprograms [TTTY78, Woo79, Poe80, TTT81, Fis81a, MPF82, IKI83, SD85]. In [Har87] Harris discussed both local and global compaction in the context of the Burroughs D-machine. Global microcode optimization under timing constraints was discussed by Su et al. in [SWX88]. The global microcode compaction techniques laid the foundation for the development of higher level program optimization techniques, viz., trace scheduling [Fis81b] (discussed in the next chapter) and the family of beyond basic block scheduling techniques. A survey of resource allocation methods for optimizing microcode compilers is available in [MDO84]. Register allocation algorithms for optimizing microcode compilers are discussed in [KT79]. In [BTT99], Bharitkar et al. proposed a new approach for optimization of microprograms based on neural networks. In [BWJ90], Beaty proposed a new approach for microcode compaction using genetic algorithms. Recently, Slechta et al. [SCF+03] proposed a dynamic approach for global optimization micro-operations.

We discuss only a few algorithms for the limited resource case; there are many more. A more in-depth treatment of Force-directed Scheduling is in [PK89]. Lookahead Scheduling [Bea92] uses *foresight* to speed up the scheduling process: at each control step it foresees the successors that might become critical after scheduling operations in the current control step. Stochastic scheduling [Sch00] (also known as randomized backward and forward scheduling) is a Monte Carlo list scheduling algorithm that breaks priority ties in a random manner, thereby trying to avoid arbitrary decisions that may lead to suboptimal results. The shortest schedule of all the runs is returned as the final schedule. Many heuristics [BR91, SB92] have been proposed for selecting the "most important" operations. One line of work has adapted ideas in machine learning and genetic programming to make the identification of the most important operations more systematic. Allan et al. [AM88] propose a *discriminating polynomial selection* procedure in which the *fitness* level of the nodes is experimentally determined (see also [Bea91, BCS96]). Another example is scheduling using the technique of *iterative repair* [CSS98] in which operations are scheduled even if there are conflicts by unscheduling the conflicting operations and then attempting to reschedule them (which, if there are additional conflicts in rescheduling may cause the unscheduling/rescheduling process to iterate); see also Section 6.6.

The references listed here constitute a small percentage of what is available.

4

TRACE SCHEDULING

Since its introduction by Joseph A. Fisher in 1979, trace scheduling has influenced much of the work on compile-time ILP. Initially developed for use in microcode compaction, trace scheduling quickly became the main technique for machine-level compile-time parallelism exploitation. Trace scheduling has been used since the 1980s in many state-of-the-art compilers (e.g., Intel, Fujitsu, HP).

The aim of this chapter is to create a mathematical theory of the foundation of trace scheduling. We give a clear algorithm showing how to insert compensation code after a trace is replaced with its schedule, and then *prove* that the resulting program is indeed equivalent to the original program. We derive an upper bound on the size of that compensation code, and show that this bound can be actually attained. We also give a very simple proof that the trace scheduling algorithm always terminates.

4.1 Introduction

Since its introduction, Trace Scheduling has influenced much of the work on compile-time ILP. Initially developed for use in microcode compaction [LDSM80], trace scheduling quickly became the main technique for machine-level compile-time parallelism exploitation. Trace scheduling has been used since the 1980's in many state-of-the-art compilers (e.g., Intel, Fujitsu, HP). With the hardware work for the pioneering VLIW (Very Large Instruction Word) machine at

A. Aiken et al., *Instruction Level Parallelism*,
DOI 10.1007/978-1-4899-7797-7_4

Multiflow Computers [LFK+93], trace scheduling has been instrumental in the development of the whole ILP area.

In this chapter, we start with a sequential program represented as a directed acyclic graph (dag), where a node stands for an assignment or a conditional branch operation, and an edge indicates control flow. The goal is to find an equivalent parallel program, also represented as a dag, where a node stands for an instruction consisting of multiple operations that can be executed in parallel (on a given machine), and an edge again indicates control flow.

We consider loop-free sequential programs that are made up of a number of basic blocks connected by conditional branches. The easiest way to parallelize such a program is to parallelize each block separately by any of the known basic-block techniques [LDSM80]. But, for better results, one must overcome the limitation to basic blocks; one needs a method that allows code to move freely (subject to dependences) across basic block boundaries. In 1979, J. A. Fisher proposed a global instruction scheduling algorithm called *Trace Scheduling* [Fis79] that does exactly that (also see [Fis81a, Fis81b]).

In a given sequential program with N conditional branch operations, there are at most 2^N distinct execution paths through the program graph. Any given input to the program selects a particular path, and only the operations lying on that path are executed. The trace scheduling algorithm starts by assigning a probability of execution to each edge and to each node in the original program graph. It picks a sequence of operations (a *trace*) lying contiguously on the most probable path through the program. The algorithm *compacts* the trace by creating a *schedule* that is a sequence of instructions equivalent to the trace. Both assignment and conditional branch operations take part in the compaction process. The trace is then replaced by its schedule. This creates a program with a parallel portion along the selected trace, but the program is still largely sequential.

This newly created program, however, may not be equivalent to the original program, since the compaction process moves operations around, and some of them would typically be missing from paths to which they originally belonged. We *compensate* by inserting copies of those operations at suitable points in the new program. The set of all such copies inserted is the *compensation code* needed for replacement of the trace in question. It can also happen that operations are added to paths where they previously did not occur; we will restrict such "extra" operations to ensure program semantics are preserved.

Once a trace has been compacted and compensation code added, another trace is picked from the remaining sequential part of the program. This trace is then compacted and compensation code is inserted to maintain program equivalence. The process is continued until the sequential part of the program is empty (or some other termination criterion is reached). There are two basic questions:

1. After replacing a trace with its schedule, exactly how do we compensate to preserve program semantics?

2. Can we guarantee that a given sequential program (of the type being considered) can always be converted into an equivalent parallel program after only a finite number of trace replacements?

The rules governing insertion of compensation code found in the existing literature tend to be ad hoc and without any formal proofs. The available proof of termination for trace scheduling is very long and involved [Nic84]. It is important to clarify the termination issue especially since there are cases where the compensation code has more operations than the trace being replaced, a situation which raises the possibility that repeatedly scheduling traces may cause the program to grow without bound and result in non-termination.

The aim of this chapter is to provide a rigorous framework in which the fundamental concepts and techniques of trace scheduling can be studied carefully. The main highlights are the following:

1. a precise algorithm describing how to create an equivalent program after replacing a trace with its schedule by inserting suitable compensation code;

2. a proof that the algorithm works, that is, the resulting program at the end of each replacement step is indeed equivalent to the program at the beginning of the step;

3. a tight upper bound on the size of the compensation code inserted after a trace replacement, along with examples where the bound is actually attained;

4. a very simple proof that the trace scheduling algorithm terminates.

In Section 4.2, we introduce the basic concepts we need. In Section 4.3, we consider replacement of traces that can be entered only at the first operation. This discussion is then generalized in Section 4.4, where we study replacement of arbitrary traces. A very broad overview of the whole trace scheduling algorithm is presented in Section 4.5, where we also provide a proof of eventual termination.

The main references on trace scheduling are [Fis79, Fis81b, Nic84, Ell85, LFK+93, Ban11b].

4.2 Basic Concepts

In this section we explain basic terms and concepts used throughout the remainder of the chapter.

4.2.1 Program Model

Recall the definition of an *operation* and an *assignment* in Section 3.2 on page 44. We add a new kind of operation, the *conditional branch operation*, which has the general form:

B: If b then go to C, else go to D

where B, C, D are labels and b is a boolean expression (the *condition*). A conditional operation evaluates b and jumps to the label C if the value of b is T (*true*), or to D if the value is F (*false*). We say $B \to C$ is the *true branch* of the conditional operation B and that $B \to D$ is the *false branch*. Such an operation does not write any memory location, but does read the locations specified in b. The conditional operation is simplified to

B: If b then go to C

when it is understood where control flows on the false branch. For example, suppose we are working with a specified sequence of operations $(A_1, A_2, \ldots, A_n)$ where some A_i (with $1 \leq i < n$) is a conditional branch. In this case, it is convenient to assume that control from A_i flows to A_{i+1} when the condition in A_i is false. A conditional branch operation is also referred to as a *test*.

If the condition in a conditional branch is always true, we omit it and get an *unconditional* branch as a special case:

B: go to C

An *instruction* for a given machine has the general form (Figure 4.1)

I: $[A_1, A_2, \ldots, A_r; B_1, B_2, \ldots, B_s]$

where **I** is a label, $A_1, A_2, \ldots, A_r$ are assignments, and $B_1, B_2, \ldots, B_s$ are tests, such that all these operations can be processed by that machine simultaneously. There are $s+1$ possible targets (not necessarily all distinct) associated with this instruction: $C_1, C_2, \ldots, C_s, D$. Each test B_i, for $1 \leq i \leq s$, has the form

B_i: If b_i then go to C_i

After all operations in the instruction have been processed, control from **I** flows to C_i if b_i is the first condition in the sequence of conditions that is true. In other words, control goes to C_i if $\bar{b}_1\bar{b}_2\cdots\bar{b}_{i-1}b_i$ is true. If all conditions are false, then control flows from **I** to some location D, where D is determined by the structure in which the instruction is placed. For example, if **I** is within a specified sequence of instructions, then D is the next instruction in the sequence.

The ordering of the assignment operations in an instruction is not important, but the ordering of the conditional branch operations is very important. For example, consider the instruction in the above

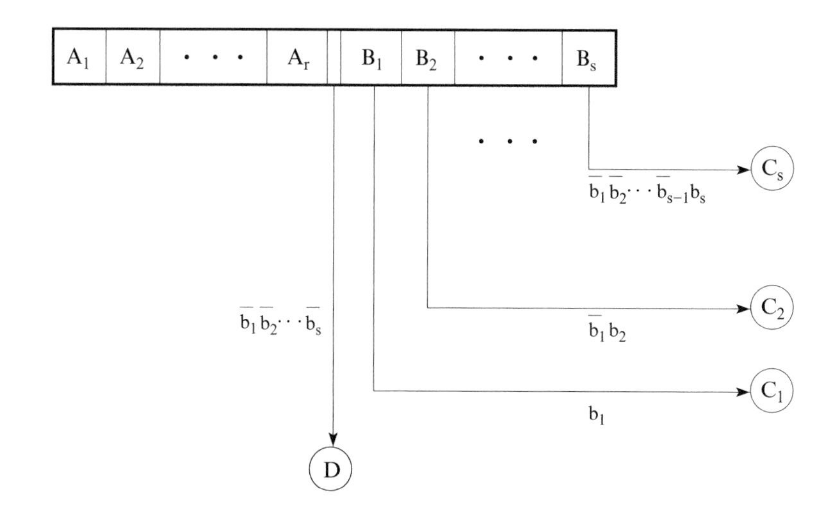

Figure 4.1 : General Form of an Instruction

paragraph, where the tests are arranged in the order $B_1, B_2, \ldots, B_s$. The predicates for the targets $C_1, C_2, \ldots, D$ are $b_1, \bar{b}_1 b_2, \ldots, \bar{b}_1 \bar{b}_2 \cdots \bar{b}_s$, respectively. If we interchange B_1 and B_2 in the instruction to get the order $B_2, B_1, \ldots, B_s$, then the predicates for $C_2, C_1, \ldots, D$ become $b_2, \bar{b}_2 b_1, \ldots, \bar{b}_2 \bar{b}_1 \cdots \bar{b}_s$, which changes the behavior of the instruction.

In practice, the number and mix of operations in an instruction depend on the resources of the machine we are compiling for. For our purposes, an instruction is any collection of operations that can be grouped together in the above form. An instruction may have an empty set of assignments, or an empty set of tests, or both.

A *program* is defined by a directed acyclic graph (dag), such that

1. there are two distinguished nodes labeled START and END;
2. every other node is an *ordinary* node that represents either a single operation or an instruction;
3. an edge represents possible flow of control from one node to another;
4. the START node has no incoming edges and the END node has no outgoing edges;
5. for each ordinary node u, there is a path from START to u and a path from u to END;
6. for each input to the program, there is a directed path from START to END through the graph.

The special nodes START and END are used as anchors for the program; they do not take part in program transformations. Unless made clear otherwise, a 'node' means an ordinary node representing an operation or an instruction.

An ordinary node may have any number of incoming edges. If an ordinary node represents an operation, then it has one outgoing edge for an assignment, and two outgoing edges for a test. An instruction node with s tests has $(s+1)$ outgoing edges.

A program has a sequential part and a parallel part: the *sequential part* consists of all the operation nodes and the edges between them, and the *parallel part* consists of all the instruction nodes and the edges between them. There are also edges going back and forth between the two parts. A program is *sequential* if its parallel part is empty. A program is *parallel* if its sequential part is empty.

For an instruction node with assignment operations $A_1, A_2, \ldots, A_r$ and tests $B_1, B_2, \ldots, B_s$, the *sequential equivalent* is the sequential program segment

$$A_1 \to A_2 \to \cdots \to A_r \to B_1 \to B_2 \to \cdots \to B_s.$$

When an input is applied to a program G, we get a definite path through G from the START node to the END node. This is the *execution path* for that input. On this path, a certain set of assignment operations are executed in a certain order leading to the output for this particular input. Two programs are *equivalent with respect to a given input* if they produce the same output for that input. Two programs are *equivalent* if they are equivalent with respect to every input.

4.2.2 Traces

By a *trace* in a program we mean a path in its sequential part. Formally, a *trace* $\mathcal{T}$ in a program G is a path of the form:

$$P \xrightarrow{e_0} A_1 \xrightarrow{e_1} A_2 \xrightarrow{e_2} \cdots \xrightarrow{e_{n-1}} A_n \xrightarrow{e_n} Q$$

where $A_1, A_2, \ldots, A_n$ are operation nodes. The *length* of $\mathcal{T}$ is n, its *anchors* are P and Q, its *on-trace operations* are $A_1, A_2, \ldots, A_n$, and its *on-trace edges* are $e_0, e_1, e_2, \ldots, e_n$. The anchors simply hold the trace in place; they do not take part in any transformation of the trace. (P could be the START node and Q the END node.) When the on-trace edges are understood, we may say that $\mathcal{T}$ is the trace

$$P \to A_1 \to A_2 \to \cdots \to A_n \to Q.$$

If the anchors are also understood, we may say that $\mathcal{T}$ is the trace $(A_1, A_2, \ldots, A_n)$.

By making trivial changes in the program, if necessary, we may assume that a trace has the following properties:

1. If A_i is a conditional branch operation on $\mathcal{T}$, then its false branch is the edge e_i, and its true branch does not go to another on-trace operation.
2. There is no incoming edge at A_1 other than e_0. For any other operation A_i on the trace, there is the incoming edge e_{i-1} from A_{i-1} on the trace, and possibly other incoming edges from nodes not on the trace.

An edge e in the program is an *off-trace* edge for $\mathcal{T}$ if one endpoint of e is an operation on the trace and the other endpoint is not. The true branch $A_\ell \to B$ of a conditional branch operation A_ℓ on the trace is an outgoing off-trace edge. An incoming off-trace edge $C \to A_p$ at an operation A_p on the trace is called a *join* at A_p. Note that there cannot be a join at A_1.

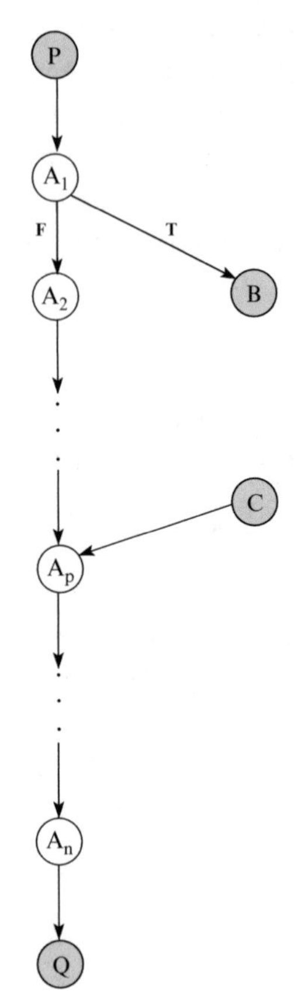

Figure 4.2 : A Typical Trace (showing a test A_1 and a join $C \to A_p$)

It helps to visualize a trace $P \to A_1 \to A_2 \to \cdots \to A_n \to Q$ as a vertical string with the operations attached to it in this order (Figure 4.2.). Then it makes sense to refer to A_1 as the 'first' or 'topmost' operation on the trace, and to A_n as the 'last' or 'lowest' operation. In this language, we can talk about an operation being 'above' or 'below' another operation (or edge), or about moving an operation 'up' or 'down' past another operation. For example, in Figure 4.2, the operations $A_p, A_{p+1}, \ldots, A_n$ are below the join $C \to A_p$.

4.2.3 Dependence

Consider a trace

$$\mathcal{T} : P \rightarrow A_1 \rightarrow A_2 \rightarrow \cdots \rightarrow A_n \rightarrow Q$$

in the sequential part of a program G. The sequential execution order of the operations implies an ordering of the various memory accesses that these operations represent which in turn creates a dependence structure; however, the notion of dependence must be extended beyond Definition 3.1 because we are no longer dealing with basic blocks but traces. Take any two operations A_i and A_j on the trace, where $1 \leq i < j \leq n$. We say A_j *depends* on A_i, and write $A_i \, \delta \, A_j$, if one of the following two conditions holds:

1. $A_i \, \delta \, A_j$ according to Definition 3.1, or
2. A_j is an assignment and A_i is a test and the location written by A_j is live on the true (i.e., off-trace) branch of A_i.

The *data dependence graph* for the trace $\mathcal{T}$ is a dag, where the nodes represent the operations on the trace and there is an edge $A_i \rightarrow A_j$ if $A_i \, \delta \, A_j$. As in Section 3.2, $\bar{\delta}$ denotes the transitive closure of δ and two operations A_i and A_j are mutually independent if $A_i \, \bar{\delta} \, A_j$ and $A_j \, \bar{\delta} \, A_i$ are both false. From here on, we simply use δ for $\bar{\delta}$.

The theory presented here is independent of the choice of machine model, programming language, or instruction set. We have given a minimal set of dependence rules that, in particular, ignores potential control flow changes due to program exceptions, which is adequate for our purposes.

The concept of dependence plays a pivotal role in program transformations. We assume the basic principle that a change in the execution order of a set of mutually independent assignment operations never changes the output of a program. For a comprehensive discussion on dependence analysis, see [Ban97].

4.2.4 Schedules

Consider a trace

$$\mathcal{T} : P \rightarrow A_1 \rightarrow A_2 \rightarrow \cdots \rightarrow A_n \rightarrow Q$$

in the sequential part of a given program G. Suppose that a permutation of operations on $\mathcal{T}$ creates the trace

$$\mathcal{T}' : P \to A_{i_1} \to A_{i_2} \to \cdots \to A_{i_n} \to Q.$$

This permutation is *valid*, if for any two operations A_r and A_s, the relation $A_r\, \delta\, A_s$ guarantees that A_r is above A_s on $\mathcal{T}'$ (i.e., $r = i_k$ and $s = i_\ell$ where $1 \le k < \ell \le n$).

A *schedule* for the trace $\mathcal{T}$ is a sequence of instructions and edges

$$P \to \mathbf{I}_1 \to \mathbf{I}_2 \to \cdots \to \mathbf{I}_m \to Q$$

with the following properties:

1. Each instruction consists of operations on the trace, and each operation on the trace appears exactly once in one of the instructions;

2. If two operations A_r and A_s on the trace are such that $A_r\, \delta\, A_s$, then A_r appears in an instruction $\mathbf{I}_j$ and A_s in an instruction $\mathbf{I}_k$ such that $j < k$;

3. For $1 \le j \le m$, if the conditions in all tests of the instruction $\mathbf{I}_j$ are false, then control flows from $\mathbf{I}_j$ to $\mathbf{I}_{j+1}$ if $j < m$, and to Q if $j = m$.

Note that property (2) implies that the operations in each instruction are pairwise mutually independent.

Compacting a trace means finding a schedule for it. After a schedule has been found for a trace, we say that the trace has been *compacted* and that the operations on the trace have been *scheduled*. Two different compaction algorithms may yield two different schedules for the same trace and the same machine.

For $1 \le j \le m$, let $\mathcal{S}_j$ denote the sequential equivalent of instruction $\mathbf{I}_j$ in the above schedule. By replacing each instruction with its sequential equivalent, we get the path

$$P \to \mathcal{S}_1 \to \mathcal{S}_2 \to \cdots \to \mathcal{S}_m \to Q$$

which is a trace of the form

$$\mathcal{T}' : P \to A_{i_1} \to A_{i_2} \to \cdots \to A_{i_n} \to Q.$$

We say this trace *generates* the schedule. Clearly, $\mathcal{T}'$ represents a valid permutation of operations on $\mathcal{T}$.

Starting from a generating trace $\mathcal{T}'$ we can easily reconstruct the schedule, by first partitioning $\mathcal{T}'$ into the sequential equivalents $\mathcal{S}_j$ of instructions $\mathbf{I}_j$, and then transforming each $\mathcal{S}_j$ back into the corresponding $\mathbf{I}_j$.

4.2.5 Program Transformation

The main idea of the trace scheduling algorithm is to increase the parallelism in a program by replacing selected traces in the sequential part by their corresponding schedules. However, for this to work, the program after replacement must be equivalent to the program before replacement. As it turns out, program equivalence does not usually happen automatically when a trace is replaced by its schedule.

Whenever a program transformation takes place, we need to ensure that the output for each input remains unchanged. If the operations on a trace are permuted and grouped together to form a schedule, some of them may end up on paths to which they do not belong, or drop out of paths to which they do belong. An extra operation may be allowed on the execution path for an input, only if that operation is dead code on that path. If an operation drops out of an execution path, a copy of that operation must be inserted on that path at a suitable point. *Copying* an operation A on an edge $B \to C$ means replacing that edge with the path $B \to A' \to C$, where A' is a copy of A. If A is a test, then $A' \to C$ would be the false branch of the test A'. (When there is no chance for confusion, we will use the same symbol for an operation and its copies.)

Thus, after a trace replacement, we may need to *compensate* the resulting program by inserting copies of several operations on the trace. The *compensation code* for a trace replacement is the set of all such copies inserted. The *size* of the compensation code is the number of operations it contains.

From a theoretical point of view, the most interesting question in trace replacement is: after a trace is replaced by its schedule, is it at all possible to compensate properly and create a program that is equivalent to the starting program? In the next two sections we answer this question affirmatively.

4.3 Traces without Joins

In this section, we study the special case of traces without joins and prepare the groundwork for the general case of arbitrary traces in the next section. The ultimate goal is to create an equivalent program after replacing a trace with its schedule. First, we study what happens when only two adjacent operations on the trace are interchanged (Theorem 4.1). Since any permutation of operations on the trace can be effected by a sequence of adjacent interchanges, we would then know how to replace a trace with a valid permutation (Theorem 4.2). Since a schedule for a trace can be built up from the permutation that generates it (Section 4.2), it is now easy to handle the situation where a trace is replaced with its schedule (Theorem 4.3). Theorem 4.4 gives a tight upper bound for the size of the compensation code needed to create an equivalent program after replacing a trace (without joins) with its schedule.

To simplify the notation in the proof of Theorem 4.1, we consider a trace with just two operations. The same result applies to a trace of any length, as long as only two adjacent operations are interchanged.

Theorem 4.1. *Consider a trace $\mathcal{T} : P \to A \to B \to Q$ without joins in the sequential part of a program G, such that B does not depend on A. Replace $\mathcal{T}$ with the trace $\mathcal{T}' : P \to B \to A \to Q$, and copy the operation A on the true branch of B in case B is a test. Then the resulting program G' is equivalent to G.*

PROOF. The programs G and G' are equivalent if they are equivalent with respect to each input. If an input is such that the execution paths for it in G and G' are the same, then the corresponding outputs are identical, and hence the two programs are equivalent with respect to that input. Thus, G and G' are already equivalent with respect to any input whose execution path in G excludes both operations A and B. For the purpose of this proof, an *interesting* input is one whose execution path in G touches at least one of A and B. Since the trace has no joins, the path of an interesting input must include the edge $P \to A$. Based on which of A and B is an assignment or a test, four possible cases arise. The four sub-figures of Figure 4.3 correspond to those cases. In each sub-figure, the graph on the left represents the part of the original program G that contains the trace $\mathcal{T}$, and the graph on the right represents the corresponding part of the

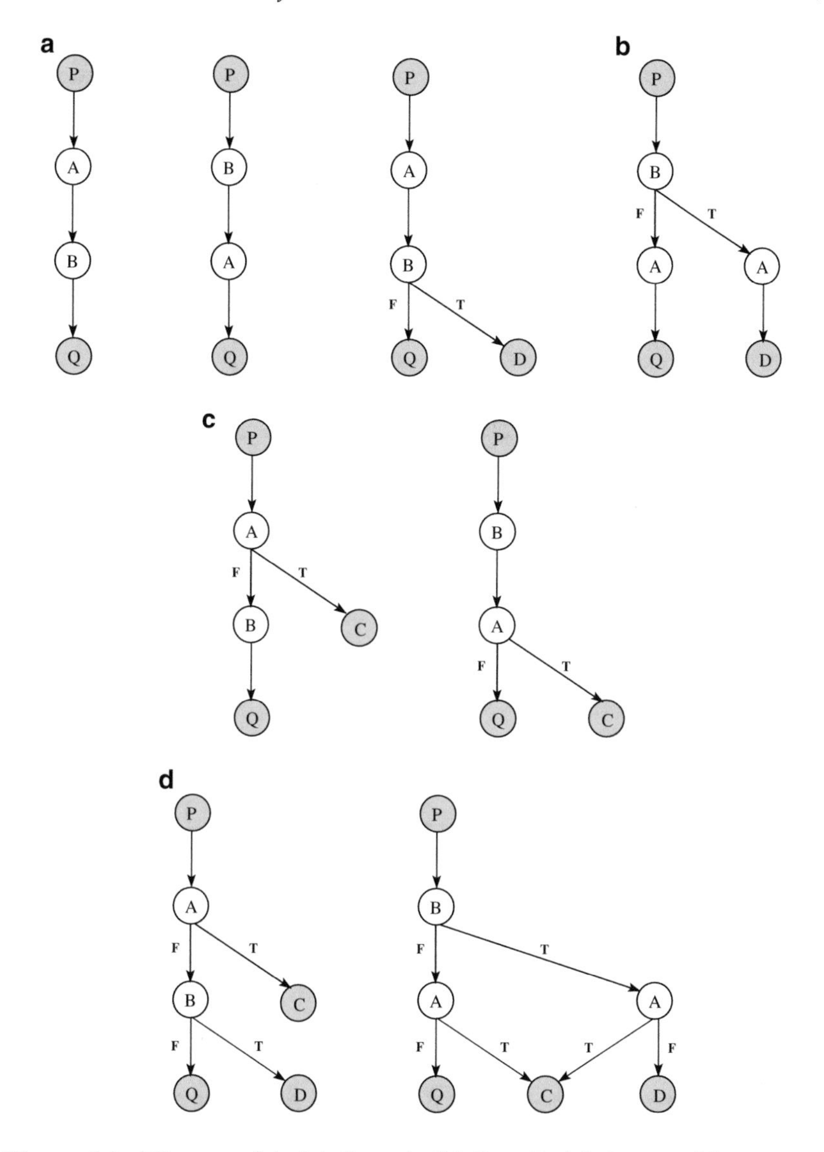

Figure 4.3 : Theorem 4.1: (a) Case A, (b) Case B, (c) Case C, (d) Case D

transformed program G'. In each case, we separate the interesting inputs into different types, and show that G and G' are equivalent with respect to any interesting input of any type.

Case A. Let both A and B be assignments (Figure 4.3(a)). For any interesting input, the execution path in G must include the segment $P \rightarrow A \rightarrow B \rightarrow Q$. The corresponding execution path in G' is different only in that it includes the segment $P \to B \to A \to Q$ instead. Since A and B are mutually independent, reversing their execution order has no effect on the output. Hence, G and G' are equivalent with respect to each interesting input.

Case B. Let A be an assignment operation and B a test (Figure 4.3(b)). Let b denote the condition and $B \to D$ the true branch of B. Since B does not depend on A, the value of the condition is the same when control comes to B from P, whether via A or directly. There are two types of interesting inputs in this case: the inputs whose execution path in G contains the segment $P \to A \to B \to Q$, and those whose path contains $P \to A \to B \to D$. For an input of the first type, we have $b = \mathrm{F}$. If such an input is applied to G', control comes to P, then goes to B, and then to A since the value of b is still F. Thus, the path in G' executes the same set of assignments in the same order as the path in G. Hence, G and G' are equivalent with respect to each input of the first type. Similarly, one can show that they are also equivalent with respect to each input of the second type.

Case C. Let operation A be a test and B an assignment (Figure 4.3(c)). Let a denote the condition and $A \to C$ the true branch of A. Since B does not depend on A, the value of the condition is the same when control comes to A from P, whether directly or via B. There are two types of interesting inputs here also: the inputs for which the execution path in G contains the segment $P \to A \to B \to Q$, and those for which the path contains $P \to A \to C$. For an input of the first type, we have $a = \mathrm{F}$. If such an input is applied to G', control comes to P, then to A via B, and then to Q since the value of a is still F. Thus, the path in G' executes the same set of assignments in the same order as the path in G. Hence, G and G' are equivalent with respect to each input of the first type. For an input of the second type ($a = \mathrm{T}$), the path in G' has one more assignment, namely B, than the path in G. This has no effect on the output, since B is necessarily dead code on the true branch of A in this case. Hence, the two programs are equivalent with respect to each input of the second type also.

Case D. Let both operations A and B be tests (Figure 4.3(d)). Let A have condition a and true branch $A \to C$, and B have condition b and true branch $B \to D$. Think of the two adjacent tests as one big switch. The structure of the switch changes as we go from G to G', but the overall function remains the same as we see below.

Take any interesting input. Its execution path in G must contain the edge $P \to A$. When this input is applied to G, control enters the switch in G at A, and then flows to C, D, or Q depending on exactly which of the following three boolean functions is true: $a, \bar{a}b, \bar{a}\bar{b}$. When it is applied to G', control enters the switch in G' at B, and then flows to C, D, or Q depending on exactly which of the following three boolean functions is true: $(ba + \bar{b}a), b\bar{a}, \bar{b}\bar{a}$. Note that the functions in this sequence are equivalent to the corresponding functions in the sequence for G. Since the values of a and b do not change from G to G', it follows that the tests in G' have the same effect as the tests in G. Thus, the outputs of G and G' must be the same for each interesting input, and therefore the two programs are equivalent.

This completes the proof of the theorem. ■

Any permutation of operations on a trace can be achieved by a finite number of adjacent interchanges. The following example illustrates the particular order of interchanges we use to prove Theorem 4.2.

Example 4.1. Consider a trace $\mathcal{T}$, and a trace $\mathcal{T}'$ that represents a permutation of operations on $\mathcal{T}$, as shown below:

$$\mathcal{T} : P \to A_1 \to A_2 \to A_3 \to A_4 \to A_5 \to Q$$
$$\mathcal{T}' : P \to A_4 \to A_3 \to A_5 \to A_2 \to A_1 \to Q.$$

It is clear that we can go from $\mathcal{T}$ to $\mathcal{T}'$ by 8 adjacent interchanges:

$$\begin{aligned}(A_1, A_2, A_3, A_4, A_5) &\Rightarrow (A_1, A_2, A_4, A_3, A_5) \Rightarrow (A_1, A_4, A_2, A_3, A_5)\\ &\Rightarrow (A_4, A_1, A_2, A_3, A_5) \Rightarrow (A_4, A_1, A_3, A_2, A_5)\\ &\Rightarrow (A_4, A_3, A_1, A_2, A_5) \Rightarrow (A_4, A_3, A_1, A_5, A_2)\\ &\Rightarrow (A_4, A_3, A_5, A_1, A_2) \Rightarrow (A_4, A_3, A_5, A_2, A_1).\end{aligned}$$

To understand the process behind this, consider the first operation A_4 on $\mathcal{T}'$. It is the fourth operation on $\mathcal{T}$. Starting with $\mathcal{T}$, we move A_4 to the first position by 3 adjacent interchanges. Now A_3, the second operation on $\mathcal{T}'$, is the fourth operation on the current trace. We move it to the second position by 2 adjacent interchanges. Then we take A_5, the third operation on $\mathcal{T}'$, which is in the fifth position on the current trace. It is brought to the third spot by 2 interchanges. Finally, A_2 and A_1 are interchanged to get the trace $\mathcal{T}'$.

In this process, an operation A_j is interchanged with a sequence of operations $A_{j_1}, A_{j_2}, \ldots, A_{j_k}$, in this order, where $j > j_1 > j_2 > \cdots > j_k \geq 1$. For example, A_5 is first interchanged with A_2 and then with A_1, where $5 > 2 > 1$.

Theorem 4.2. *Consider a trace*

$$\mathcal{T} : P \to A_1 \to A_2 \to \cdots \to A_n \to Q$$

without joins in the sequential part of a program G. Replace $\mathcal{T}$ with a trace

$$\mathcal{T}' : P \to A_{i_1} \to A_{i_2} \to \cdots \to A_{i_n} \to Q$$

that represents a valid permutation of the operations on $\mathcal{T}$. Copy on to the true branch $A_\ell \to B$ of each test A_ℓ of trace $\mathcal{T}'$ all operations that were above A_ℓ on $\mathcal{T}$ and are now below A_ℓ on $\mathcal{T}'$. The order of copied operations from A_ℓ to B is the original order on trace $\mathcal{T}$. The resulting program G' is equivalent to the original program G.

PROOF. We accomplish the transformation $\mathcal{T} \Rightarrow \mathcal{T}'$ in $(n-1)$ steps, where in each step we move an operation to its position on $\mathcal{T}'$ through a sequence of adjacent interchanges (as described in Example 4.1). Theorem 4.1 is invoked after each interchange.

In Step 1, we take the operation A_{i_1}, the first operation on $\mathcal{T}'$. There are $(i_1 - 1)$ operations on top of it on $\mathcal{T}$. Clearly, A_{i_1} cannot depend on any one of them, since the permutation that moves A_{i_1} to the first place on $\mathcal{T}'$ is valid (by hypothesis). Starting with the trace $\mathcal{T}$, we can bring A_{i_1} to the first place by interchanging it with the operations $A_{i_1-1}, A_{i_1-2}, \ldots, A_1$, one by one, in this order. If A_{i_1} is an assignment operation, then after each interchange the program remains equivalent to G (Theorem 4.1, Cases **A** and **C**). If A_{i_1} is a test, then after each interchange we copy the other operation on the true branch of A_{i_1} to keep the program equivalent to G (Theorem 4.1, Cases **B** and **D**). After $(i_1 - 1)$ adjacent interchanges, we have moved A_{i_1} to the first position, and $\mathcal{T}$ has permuted into the trace

$$\mathcal{T}_1 : P \to A_{i_1} \to A_1 \to \cdots \to A_{i_1-1} \to A_{i_1+1} \to \cdots \to A_n \to Q.$$

If A_{i_1} is a test with an original true branch $A_{i_1} \to B$, then for the A_{i_1} on $\mathcal{T}_1$ that edge has been replaced with the path

$$A_{i_1} \to A_1 \to A_2 \to \cdots \to A_{i_1-1} \to B.$$

Thus, we have copied on the true branch of A_{i_1} on $\mathcal{T}_1$, all operations that were above A_{i_1} on $\mathcal{T}$ and are now below it on $\mathcal{T}_1$ (and will be on $\mathcal{T}'$), in their original order (looking from A_{i_1} to B).

In Step 2, we start with the trace $\mathcal{T}_1$ and locate the operation A_{i_2}. Between A_{i_1} and A_{i_2} on this trace, we have the operations in the set

$$\{A_1, A_2, \ldots, A_{i_2-1}\} - \{A_{i_1}\}.$$

Clearly, A_{i_2} does not depend on any one of them. We can bring A_{i_2} to the second place by interchanging it with these operations one by one, in the order of decreasing subscripts. As before, if A_{i_2} is an assignment operation, then after each interchange the program remains equivalent to G. If A_{i_2} is a test, then after each interchange we copy the other operation on the true branch of A_{i_2} to keep the program equivalent to G. After $(i_2 - 1)$ or $(i_2 - 2)$ adjacent interchanges, we have moved A_{i_2} to the second position, and $\mathcal{T}_1$ has permuted into a trace of the form

$$\mathcal{T}_2 : P \to A_{i_1} \to A_{i_2} \to A_1 \to \cdots \to A_n \to Q,$$

where A_{i_1} and A_{i_2} are missing on the path from A_1 to A_n. If A_{i_2} is a test with an original true branch $A_{i_2} \to C$, then for the A_{i_2} on $\mathcal{T}_2$, we have copied on its true branch, all operations that were above A_{i_2} on $\mathcal{T}_1$ (and on $\mathcal{T}$) and are below it on $\mathcal{T}_2$ (and will be on $\mathcal{T}'$), in their original order (looking from A_{i_2} to C).

It is now clear how to move the operations $A_{i_3}, A_{i_4}, \ldots, A_{i_{n-1}}$ to their respective positions on $\mathcal{T}'$. And after all those moves have been made, A_{i_n} would already be in the last spot. The program that we get at the end is exactly G' and it is equivalent to G. ■

The program segment in the following example is taken from the example on page 61 of [Ell85].

Example 4.2. Consider the program segment

```
P:  ...
A:  if e1 then go to D
B:  i = i + 1
C:  if e2 then go to E
Q:  ...
```

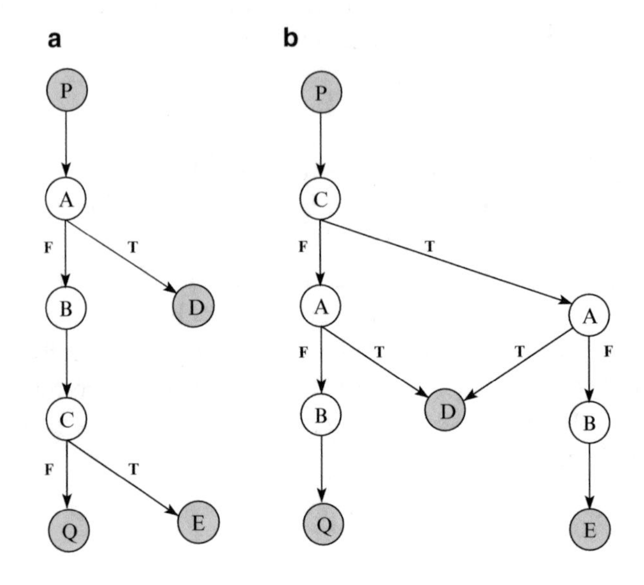

Figure 4.4 : (a) Program G, (b) Program G′

(Note that operation labels have been changed.) The graph G representing this segment is shown in Figure 4.4(a). Consider the trace

$$\mathcal{T} : P \to A \to B \to C \to Q.$$

There is one assignment operation B, and two tests A and C with true branches $A \to D$ and $C \to E$, respectively.

Assume that the trace

$$\mathcal{T}' : P \to C \to A \to B \to Q$$

represents a valid permutation of the operations on $\mathcal{T}$. Let us replace $\mathcal{T}$ with $\mathcal{T}'$. The operations A and B were above C on $\mathcal{T}$ and are now below C on $\mathcal{T}'$. So, we copy A and B on the true branch $C \to E$

of C, in the order of their appearance on $\mathcal{T}$. Since no operation has gone down the test A in this permutation, nothing needs to be copied on the true branch $A \to D$ of A. The resulting program G' of Figure 4.4(b) is equivalent to the original program G by Theorem 4.2.

It is now straightforward to replace a trace with its schedule, as we see in the following theorem.

Theorem 4.3. *Consider a trace*

$$\mathcal{T} : P \to A_1 \to A_2 \to \cdots \to A_n \to Q$$

without joins in the sequential part of a program G. Replace $\mathcal{T}$ with a schedule for $\mathcal{T}$

$$P \to \mathbf{I}_1 \to \mathbf{I}_2 \to \cdots \to \mathbf{I}_m \to Q.$$

Copy on the true branch $A_\ell \to B$ of each test A_ℓ in each instruction $\mathbf{I}_k$, all operations in the set $\{A_1, A_2, \ldots, A_{\ell-1}\}$ that are either in $\mathbf{I}_k$ to the right of A_ℓ, or in instructions $\mathbf{I}_{k+1}, \mathbf{I}_{k+2}, \ldots, \mathbf{I}_m$. (The order of copied operations from A_ℓ to B is to match their original order on $\mathcal{T}$.) Then the resulting program G' is equivalent to the original program G.

PROOF. Let $\mathcal{T}'$ denote the trace that generates the given schedule (Section 4.2). Then $\mathcal{T}'$ represents a valid permutation of operations on $\mathcal{T}$. Replace the trace $\mathcal{T}$ in G with the trace $\mathcal{T}'$. Then, copy on the true branch $A_\ell \to B$ of each test A_ℓ on $\mathcal{T}'$, all operations that were above A_ℓ on $\mathcal{T}$ and are now below it on $\mathcal{T}'$. These operations come from the set $\{A_1, A_2, \ldots, A_{\ell-1}\}$, and they are to be copied in their original order (i.e., in the order of increasing subscripts) from A_ℓ to B. Let G_1 denote the resulting program. By Theorem 4.2, G_1 is equivalent to G.

Next, partition $\mathcal{T}'$ into the sequential equivalents $\mathcal{S}_1, \mathcal{S}_2, \ldots, \mathcal{S}_m$ of the instructions $\mathbf{I}_1, \mathbf{I}_2, \ldots, \mathbf{I}_m$ to get the path

$$P \to \mathcal{S}_1 \to \mathcal{S}_2 \to \cdots \to \mathcal{S}_m \to Q.$$

Then consider replacing each segment $\mathcal{S}_j$ with the corresponding instruction $\mathbf{I}_j$. Each $\mathcal{S}_j$ consists of a sequence of assignments followed by a sequence of tests. Since these assignments are pairwise mutually independent, changing their execution order from sequential in

$\mathcal{S}_j$ to parallel in $\mathbf{I}_j$ has no effect on the output of the program. The conditions in the tests are independent of the results computed by the assignments. The effect of executing the tests in the instruction $\mathbf{I}_j$ is the same as the effect of executing them sequentially as part of $\mathcal{S}_j$. (See the interpretation of multiple tests in an instruction in Section 4.2.) Thus, the program obtained by replacing each $\mathcal{S}_j$ in G_1 with the corresponding $\mathbf{I}_j$, is equivalent to G_1. But, that program is precisely the program G'. Hence, G' is equivalent to G_1, and therefore to G. ■

Next, we find an upper bound on the size of the compensation code inserted after a trace compaction.

Theorem 4.4. *The size of compensation code inserted after replacing a trace (without joins) by its schedule, cannot exceed $t(2n - t - 1)/2$, where n is the length of the trace and t is the number of tests on it.*

PROOF. It is clear from the proof of Theorem 4.3 that we can focus on the compensation code inserted in Theorem 4.2, after replacing a trace with a valid permutation. Once a trace is replaced with the trace that generates the schedule and the program has been properly compensated, no more compensation is needed to go from the generating trace to the schedule itself.

Let

$$\mathcal{T} : P \to A_1 \to A_2 \to \cdots \to A_n \to Q$$

denote a trace without joins that has n operations including t tests and let $\mathcal{T}'$ be a valid permutation. Let k_1 denote the number of operations copied on the true branch of the highest test on $\mathcal{T}'$, k_2 the number of operations copied on the true branch of the second highest test on $\mathcal{T}'$, and so on. Then the size of the compensation code is $(k_1 + k_2 + \cdots + k_t)$.

An operation is copied on the true branch of a test, if it was above the test on $\mathcal{T}$ and is now below it on $\mathcal{T}'$, that is, if it has gone down the test in the transformation $\mathcal{T} \Rightarrow \mathcal{T}'$. Since n is the total number of operations on the trace, the maximum number of operations that can go down a test is $(n - 1)$. This happens if A_n is a test and if it is the first operation on $\mathcal{T}'$. Assume this is so. Then the maximum number of operations that can go down another test is $(n - 2)$. This happens if A_{n-1} is a test and if it is the second operation on $\mathcal{T}'$. Continuing this way, we see that the size of the compensation code is maximized when the operations $A_1, A_2, \ldots, A_{n-t}$ are assignments, the operations $A_{n-t+1}, A_{n-t+2}, \ldots, A_n$ are tests, and $\mathcal{T}'$ has the form:

$$\mathcal{T}' : P \to A_n \to A_{n-1} \to \cdots \to A_{n-t+1} \to A_1 \to A_2 \to \cdots \to A_{n-t} \to Q.$$

The maximum possible value of the size of the compensation code is then

$$(n-1) + (n-2) + \cdots + (n-t) = tn - t(t+1)/2 = t(2n-t-1)/2. \blacksquare$$

It is clear from the proof of Theorem 4.4 that the upper bound on the compensation code can actually be attained if the t tests are positioned properly on the traces. Consider the program shown in Figure 4.4. Here we have $n = 3$ and $t = 2$. According to Theorem 4.4, the number of operations in the compensation code cannot exceed $2(2*3 - 2 - 1)/2$ or 3. The actual compensation code here consists of two operations: a copy of A and a copy of B.

Corollary 1. *For a trace of length n with any number of tests, the size of the compensation code cannot exceed $n(n-1)/2$.*

PROOF. If the trace has t tests, then the size of the compensation code cannot exceed $t(2n - t - 1)/2$. Keep n fixed, and consider a real-valued function f of a real variable t, defined by

$$f(t) = t(2n - t - 1)/2 = (n - 1/2)t - t^2/2.$$

Since $f'(t) = (n - 1/2) - t$, it follows that $f'(t) > 0$ for $t < n - 1/2$ and $f'(t) < 0$ for $t > n - 1/2$. Thus, f increases in the interval $(-\infty, n - \frac{1}{2}]$ and decreases in the interval $[n - \frac{1}{2}, \infty)$. When t is an integer, the maximum value of $f(t)$ is the larger of the two values $f(t_1)$ and $f(t_2)$, where $t_1 = \lfloor n - \frac{1}{2} \rfloor = n - 1$ and $t_2 = \lceil n - \frac{1}{2} \rceil = n$. Since

$$f(t_1) = f(n-1) = n(n-1)/2 = f(n) = f(t_2),$$

the maximum value of $f(t)$ (for integral values of t) is $n(n-1)/2$. The size of the compensation code cannot exceed this value, irrespective of the number of tests on the trace. $\blacksquare$

Remark 4.1. The bound is tight. Consider a trace of the form $(A_1, A_2, \ldots, A_n)$, where A_1 is either an assignment or a test, and $A_2, A_3, \ldots, A_n$ are tests. Suppose that it is transformed into the trace $(A_n, A_{n-1},$

$\ldots, A_1$). Then, $(n-1)$ operations have gone down the test A_n, $(n-2)$ operations down A_{n-1}, and so on. Copies of those operations constitute the compensation code, and their total number is

$$(n-1) + (n-2) + \cdots + 1 = n(n-1)/2.$$

Thus, the size of the compensation code is exactly $n(n-1)/2$. When the first operation A_1 is an assignment, there are $(n-1)$ tests; this corresponds to the case $t = t_1$ in the above corollary. When A_1 is a test, we get the case $t = t_2$.

Remark 4.2. In the example of Remark 4.1, the size of the compensation code is $n(n-1)/2$, while the number of operations on the trace is n. Now $n(n-1)/2 - n = n(n-3)/2 > 0$, if $n > 3$. Thus, it is possible to have a situation where we replace a trace with a valid permutation (or a schedule), and then end up inserting more copies of operations as compensation than there are operations on the trace.

4.4 General Traces

We are now ready to handle a general trace with any number of tests and joins. Algorithm 4.1 describes how to replace such a trace with its schedule, and then insert appropriate compensation code to create an equivalent program. Theorem 4.5 proves that the algorithm works.

Algorithm 4.1. Assume that we are given a program G, a trace

$$\mathcal{T} : P \to A_1 \to A_2 \to \cdots \to A_n \to Q$$

in the sequential part of G, and a schedule

$$P \to \mathbf{I}_1 \to \mathbf{I}_2 \to \cdots \to \mathbf{I}_m \to Q$$

for $\mathcal{T}$. This algorithm describes how to replace the trace with its schedule, and then insert suitable compensation code to create a program G' equivalent to G.

1. Each join of $\mathcal{T}$ has the form $C \to A_p$, where $2 \leq p \leq n$. Replace it with the edge $C \to Q$, and then copy on that edge the operations $A_p, A_{p+1}, \ldots, A_n$ to create a trace of the form

   $$\mathcal{S} : C \to A_p \to A_{p+1} \to \cdots \to A_n \to Q.$$

   The trace $\mathcal{T}$ in the current program is now without any joins.

2. (a) Replace $\mathcal{T}$ with its schedule.
   (b) For each instruction $\mathbf{I}_k$ in the schedule and each test A_ℓ in that instruction, find all operations in the set $\{A_1, A_2, \ldots, A_{\ell-1}\}$ that are now either in $\mathbf{I}_k$ to the right of A_ℓ, or in instructions $\mathbf{I}_{k+1}, \mathbf{I}_{k+2}, \ldots, \mathbf{I}_m$. On the true branch of A_ℓ, copy those operations in their original order on $\mathcal{T}$ (in the direction from A_ℓ).

3. For each trace

$$\mathcal{S} : C \to A_p \to A_{p+1} \to \cdots \to A_n \to Q$$

created in Step 1, execute the following steps:

   (a) Find the unique integer q in $1 \leq q \leq m$, such that the instruction $\mathbf{I}_q$ contains at least one operation from the set $\{A_1, A_2, \ldots, A_{p-1}\}$, but all the instructions $\mathbf{I}_{q+1}, \mathbf{I}_{q+2}, \ldots, \mathbf{I}_m$ contain only operations from the set $\{A_p, A_{p+1}, \ldots, A_n\}$.
   (b) Delete from $\mathcal{S}$ all operations that are copies of operations in the instructions $\mathbf{I}_{q+1}, \mathbf{I}_{q+2}, \ldots, \mathbf{I}_m$. Let

$$\mathcal{S}' : C \to A_{i_1} \to A_{i_2} \to \cdots \to A_{i_r} \to Q$$

   denote the resulting trace, where $p \leq i_1 < i_2 < \cdots < i_r \leq n$.
   (c) Replace the edge $A_{i_r} \to Q$ with the edge $A_{i_r} \to \mathbf{I}_{q+1}$, if $q < m$.
   (d) On the true branch of each test A_ℓ on $\mathcal{S}'$, copy (in their original order) all operations from the set $\{A_p, A_{p+1}, \ldots, A_{\ell-1}\}$ that are in instructions $\mathbf{I}_{q+1}$ through $\mathbf{I}_m$.

4. Let G' denote the program obtained after Step 3.

Theorem 4.5. *The program G' created in Algorithm 4.1 is equivalent to the original program G.*

PROOF. Let G_1 denote the program that results at the end of Step 1. Clearly, G_1 is equivalent to G, since for each given input the execution paths through G and G_1 are identical.

At the start of Step 2, the trace $\mathcal{T}$ in the program G_1 is without any joins. It is replaced with its schedule in Step 2(a), and then in Step 2(b) compensation code is inserted as required by Theorem 4.3. Let G_2 denote the program that results at the end of Step 2. By Theorem 4.3, G_2 is equivalent to G_1.

In Step 3, we start with the program G_2 and process the traces $\mathcal{S}$. Note that each such trace $\mathcal{S}$ is without joins, and that the sequence

$$C \to A_{i_1} \to A_{i_2} \to \cdots \to A_{i_r} \to \mathbf{I}_{q+1} \to \mathbf{I}_{q+2} \to \cdots \to \mathbf{I}_m \to Q$$

is a schedule for $\mathcal{S}$. At the end of Step 3(c), we have replaced $\mathcal{S}$ with its schedule. In Step 3(d), we copy operations on the true branches of the tests in the initial part

$$C \to A_{i_1} \to A_{i_2} \to \cdots \to A_{i_r}$$

of the schedule, as mandated in Theorem 4.2. During execution of Step 2(b), operations have already been copied on the true branches of the tests in the final part

$$\mathbf{I}_{q+1} \to \mathbf{I}_{q+2} \to \cdots \to \mathbf{I}_m \to Q$$

as required by Theorem 4.3. It follows that the program G' at the end of Step 3 is equivalent to the program G_2. Since G_2 is equivalent to G_1 and G_1 to G, we see that G' is equivalent to G. ■

Example 4.3. Consider the program segment G defined by the graph in Figure 4.5(a). Focus on the trace

$$\mathcal{T} : P \to A \to B \to C \to D \to E \to Q.$$

There are two tests B and D with true branches $B \to K$ and $D \to J$, and one join $H \to C$.

Let $(\mathbf{I}_1, \mathbf{I}_2, \mathbf{I}_3)$ denote a schedule for $\mathcal{T}$, where the first instruction $\mathbf{I}_1$ consists of a single operation B, $\mathbf{I}_2$ has the operations A and D, and $\mathbf{I}_3$ has C and E. Let us apply Algorithm 4.1 to the program of this example.

1. Replace the join $H \to C$ with the edge $H \to Q$, and then copy the operations C, D, E on it to get the trace $\mathcal{S} : H \to C \to D \to E \to Q$. This creates the program G_1 in the proof of Theorem 4.5. See Figure 4.5(b).

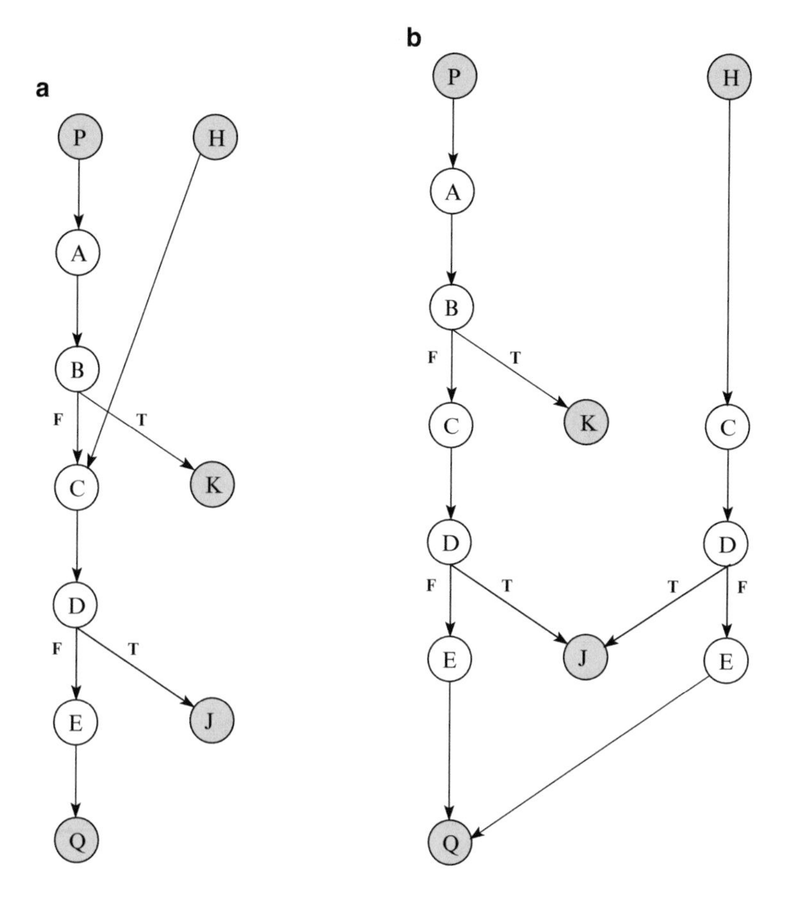

Figure 4.5 : (a) Original Program G, (b) Program G_1

2. (a) Replace the trace $\mathcal{T}$ with its schedule $(\mathbf{I}_1, \mathbf{I}_2, \mathbf{I}_3)$.
   (b) The instruction $\mathbf{I}_1$ has a test B. The only operation that was above B on $\mathcal{T}$ and is now in a lower instruction (namely $\mathbf{I}_2$) is A. Insert a copy of A on the true branch of B. The instruction $\mathbf{I}_2$ has a test D. The only operation that was above D on $\mathcal{T}$ and is now in a lower instruction (namely $\mathbf{I}_3$) is C. Insert a copy of C on the true branch of D.

   We now have the program G_2 in the proof of Theorem 4.5. See Figure 4.6(a).

3. The only trace created in Step 1 is $\mathcal{S} : H \to C \to D \to E \to Q$.
   (a) $q = 2$ is the unique integer in $1 \leq q \leq 3$, such that the instruction $\mathbf{I}_2$ has at least one operation (namely A) that was

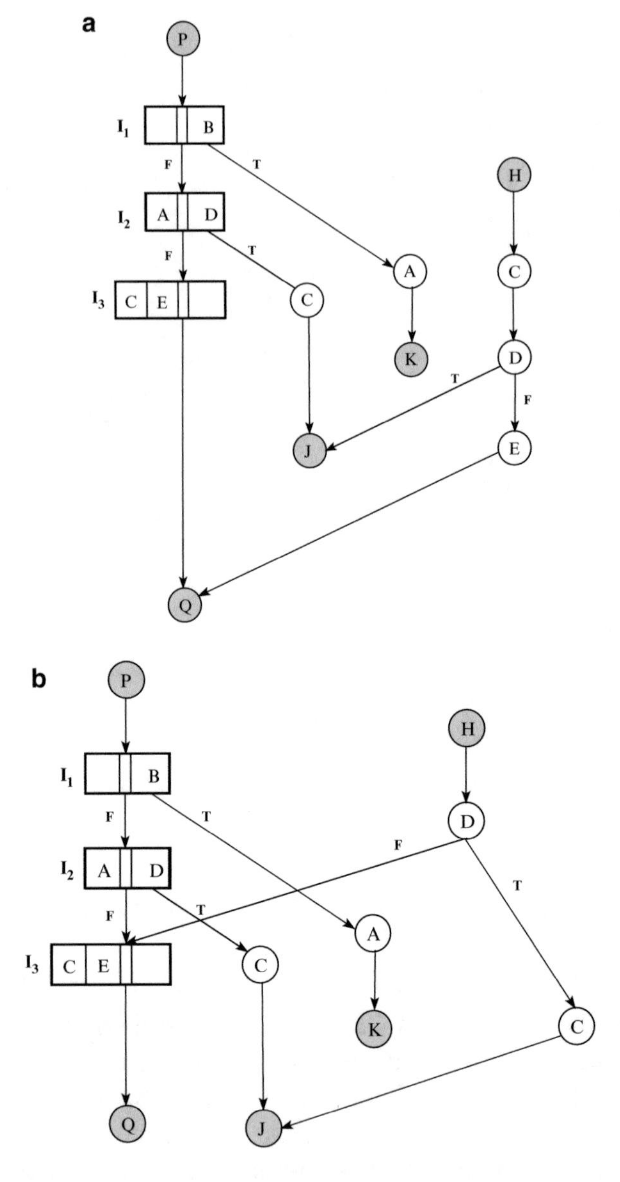

Figure 4.6 : (a) Program G_2, (b) Final Program G'

above the join $H \to C$ on the trace $\mathcal{T}$, and the instruction $\mathbf{I}_3$ contains only operations below the join.

(b) Delete from $\mathcal{S}$ all operations that are copies of operations in the instruction $\mathbf{I}_3$, to get the trace $\mathcal{S}' : H \to D \to Q$.

(c) Replace the edge $D \to Q$ with the edge $D \to \mathbf{I}_3$.

(d) The only test on $\mathcal{S}'$ is D. The only operation that was above D on the trace $\mathcal{S}$ and is now in the instruction $\mathbf{I}_3$, is C. Insert a copy of C on the true branch of D.

4. We now have the program G' of Theorem 4.5. See Figure 4.6(b).

The program G' of Figure 4.6(b) is equivalent to the program G of Figure 4.5(a).

Next, we give an upper bound for the size of compensation code needed for a single join on a trace, when it is replaced by its schedule, as described in Algorithm 4.1.

Theorem 4.6. *If b denotes the number of operations below a join on the trace $\mathcal{T}$ in Algorithm 4.1, then the size of compensation code needed to process that join cannot exceed $\lfloor (b+1)^2/4 \rfloor$.*

PROOF. We use the notation of Algorithm 4.1. Consider a join $C \to A_p$ on the trace $\mathcal{T}$, and let b denote the number of operations below the join (i.e., $b = n - p + 1$).

At the end of Step 1, the compensation code for this join consists of copies of operations $A_p, A_{p+1}, \ldots, A_n$. This code remains unchanged at the end of Step 2.

At the end of Step 3(b), the compensation code consists of copies of r operations: $A_{i_1}, A_{i_2}, \ldots, A_{i_r}$. The $(b-r)$ operations whose copies have been deleted from $\mathcal{S}$ are precisely the operations that constitute the instructions $\mathbf{I}_{q+1}, \mathbf{I}_{q+2}, \ldots, \mathbf{I}_m$.

In Step 3(d), the compensation code for this join grows again. On the true branch of each test left on $\mathcal{S}'$, we copy operations that were above the test on the trace $\mathcal{S}$, but are now in instructions $\mathbf{I}_{q+1}, \mathbf{I}_{q+2}, \ldots, \mathbf{I}_m$. Since the total number of operations in these instructions is $(b-r)$, the number of operations copied on the true branch of a test cannot exceed $(b-r)$. Since $\mathcal{S}'$ can have at most r tests, the total number of copies inserted in Step 3(d) (for the join under consideration) cannot exceed $r(b-r)$. Counting in the r operations on $\mathcal{S}'$, we see that an upper bound for the size of compensation code for the join is

$$r + r(b-r) = r(b+1-r) = (b+1)r - r^2.$$

The next step is to find the maximum value of this upper bound as r varies between $0 \le r \le b$. Allow r to have real values, and consider the real-valued function

$$h(r) = (b+1)r - r^2, \qquad -\infty < r < \infty.$$

The derivative of h with respect to r is $h'(r) = (b+1) - 2r$. Since $h'(r) > 0$ for $r < (b+1)/2$, and $h'(r) < 0$ for $r > (b+1)/2$, it follows that h is increasing in the interval $(-\infty, (b+1)/2]$ and decreasing in the interval $[(b+1)/2, \infty)$. Thus, when r is restricted to integers, the maximum value of $h(r)$ is the larger of the two values $h(r_1)$ and $h(r_2)$, where $r_1 = \lfloor (b+1)/2 \rfloor$ and $r_2 = \lceil (b+1)/2 \rceil$. In fact, it turns out that $h(r_1) = h(r_2)$.

First, let b be an odd integer. Then, $(b+1)/2$ is an integer, and $r_1 = r_2 = (b+1)/2$. The maximum value of $h(r)$ in this case is

$$h(r_1) = h(r_2) = (b+1)^2/4 = \lfloor (b+1)^2/4 \rfloor.$$

Next, let b be an even integer. Then, $(b+1)/2$ is not an integer, and we have $r_1 = b/2$ and $r_2 = b/2 + 1$. The maximum value of $h(r)$ in this case is

$$h(r_1) = h(r_2) = (b/2)(b/2+1) = \lfloor b^2/4 + b/2 + 1/4 \rfloor = \lfloor (b+1)^2/4 \rfloor.$$

Thus, the value of the upper bound on the size of compensation code can never exceed $\lfloor (b+1)^2/4 \rfloor$. This completes the proof. ■

Remark 4.3. The bound in Theorem 4.6 is tight. Consider the trace

$$\mathcal{T} : P \to A \to B \to C \to D \to E \to Q$$

in the program segment G represented by the graph in Figure 4.7. It has two tests D and E, and a join $H \to B$. Suppose

$$\mathcal{T}' : P \to D \to E \to A \to C \to B \to Q$$

is a valid permutation of $\mathcal{T}$. We can think of $\mathcal{T}'$ as a degenerate schedule for $\mathcal{T}$, where each instruction has only one operation. We get the equivalent program G' shown in Figure 4.8, after replacing $\mathcal{T}$ with $\mathcal{T}'$ and then inserting compensation code according to Algorithm 4.1. Now, the number of operations below the join $H \to B$ in G is $b = 4$. According to Theorem 4.6, the size of compensation code for this join cannot exceed $\lfloor (b+1)^2/4 \rfloor = \lfloor 25/4 \rfloor = 6$. This is precisely the number of copies of operations that have been inserted in Figure 4.8 to handle the join.

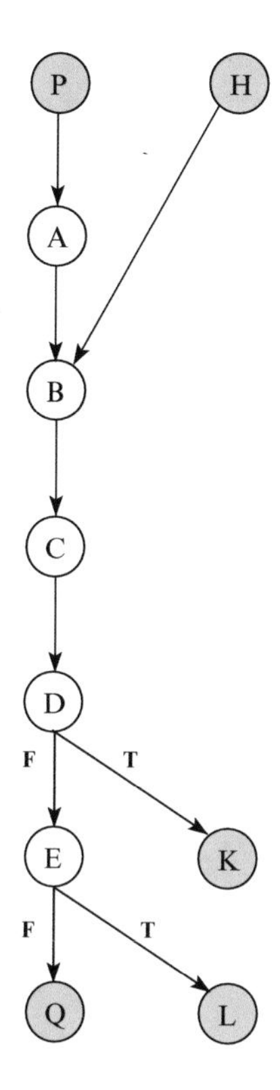

Figure 4.7 : Original Program G

Note that in Algorithm 4.1, the compensation codes for different joins are inserted independently of one another, and independently of the code inserted for the tests on the original trace $\mathcal{T}$. Hence, we can combine the results in Theorems 4 and 6, and get the following final result on total compensation code.

Theorem 4.7. *Consider a trace $\mathcal{T}$ of length n in a program G. Let there be t tests and j joins on this trace. Let there be $b_1, b_2, \ldots, b_j$ on-trace operations below those joins. Then the size of the total compensation code for replacing $\mathcal{T}$ with its schedule by Algorithm 4.1 cannot exceed*

$$\frac{t(2n-t-1)}{2} + \sum_{k=1}^{j} \left\lfloor \frac{(b_k+1)^2}{4} \right\rfloor .$$

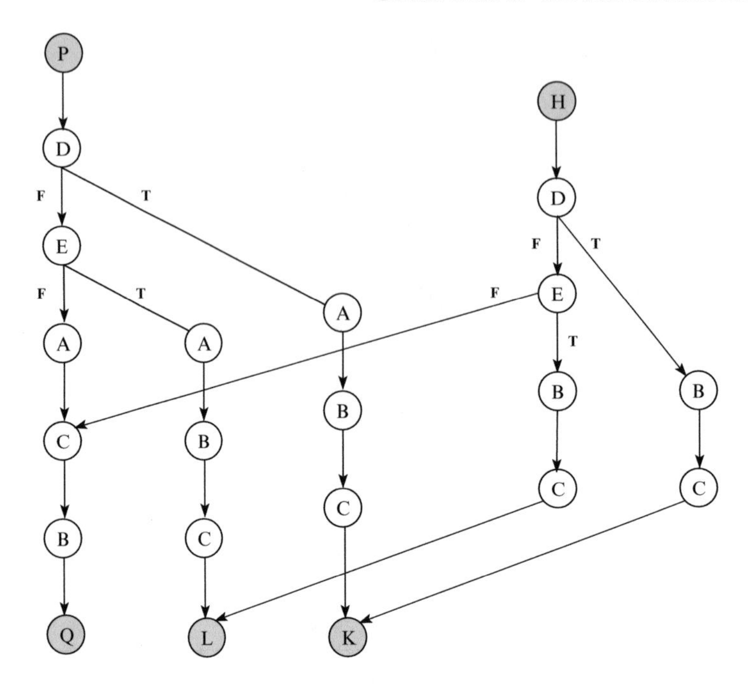

Figure 4.8 : Final Program G′

Remark 4.4. We will continue using here the notation used in Algorithm 4.1 and in the proof of Theorem 4.5. Consider a typical join $C \to A_p$ on the trace. As before, $b = n - p + 1$ denotes the number of operations below this join. At the end of Step 1, the size of compensation code for this join is b. This remains unchanged at the end of Step 2. Thus, if we quit before Step 3, we would have a program G_2 equivalent to G and a compensation code (for this join) of size b.

Now, Step 3 changes the size of the compensation code to $(r + R)$, where r is the number of operations on the trace S', and R is the number of copies inserted in Step 3(d). This number could be as low as zero, or as high as $\lfloor (b+1)^2/4 \rfloor$ by Theorem 4.6. Thus, if we want to minimize the size of compensation code for each join, we could slightly modify Algorithm 4.1 by adding an extra sub-step in Step 3:

(e) If the total number of copies of operations inserted in Step 3(d) is more than $(b - r)$, then undo steps 3(a)–3(d).

Remark 4.5. *Superblock Scheduling,* introduced by Hwu et al. [HMC+93], is a variation on Trace Scheduling that uses a truncated version of Algorithm 4.1. A *superblock* is defined as a trace without any joins. The process of removing joins from a trace is called *tail duplication*. (Step 1 of Algorithm 4.1 removes joins temporarily.) In Superblock Scheduling, we *always* skip Step 3, and therefore the size of compensation code for a join is always b. Thus, for a given join, the size of compensation code in Superblock Scheduling may be more or less than that in the unmodified Algorithm 4.1.

Remark 4.6. *Hyperblock Scheduling*, introduced by Mahlke et al. [MLC+92], is a variation of superblock scheduling. A *hyperblock* is a set of predicated basic blocks in which control may only enter from the top, but may exit from one or more locations. Hyperblocks combine basic blocks from multiple paths of control and are formed using a modified version of if-conversion [Tow76, AKPW83, TLS90]. Basic blocks are included in a hyperblock based on their execution frequency, size, and instruction characteristics. Speculative execution is provided by performing predicate promotion within a hyperblock. Superblock Scheduling often performs better than Hyperblock Scheduling for lower issue rates due to the lack of available instruction slots to schedule the instructions from the multiple paths of control. On the other hand, Hyperblock Scheduling generally provides performance for higher issue rate processors since there are a greater number of independent instructions from the multiple paths of control to fill the available processor resources.

4.5 Trace Scheduling Algorithm

The *Trace Scheduling Algorithm* acts on a sequential program and transforms it into an equivalent parallel program.[1] This transformation is achieved incrementally over a finite sequence of similar steps. In a typical step, we start with a program G (with a sequential and a parallel part) equivalent to the input program, and end with an equivalent program G' with a potentially larger parallel part. Although the sequential part of G' may not always be smaller than the sequential part of G, it is guaranteed that the sequential part of the original

[1]For simplicity, we assume that we are starting with a completely sequential program and ending with an equivalent completely parallel program. In practice, we can always start with a sequential or a partially parallel program and stop when we have an equivalent program that is sufficiently more parallel (based on some criteria) than the input program.

input program will eventually be reduced to zero in a finite sequence of steps.

A very high-level description of the algorithm is given below.

Algorithm 4.2. (Trace Scheduling). The input to this algorithm is a sequential program G_{seq}, and the output is an equivalent parallel program G_{par}.

1. Set $G \leftarrow G_{\text{seq}}$.
2. Pick a trace $\mathcal{T}$ in the sequential part of G using a set of heuristics previously agreed upon.
3. Construct the data dependence graph of $\mathcal{T}$.
4. Find a schedule for $\mathcal{T}$ with respect to the given machine, using a suitable basic-block technique.
5. Replace $\mathcal{T}$ in G with its schedule and then insert suitable compensation code to create a program G' equivalent to G (Algorithm 4.1).
6. If the sequential part of G' is nonempty, then set $G \leftarrow G'$ and go to Step 2.
7. Set $G_{\text{par}} \leftarrow G'$.

Because each iteration of steps 2–6 preserves program semantics, it follows that the program at the end of Step 6 is always equivalent to the original program G_{seq}. So, if the algorithm eventually stops, we will end with a parallel program equivalent to the original sequential program. As we see below, Algorithm 4.2 does indeed terminate.

Theorem 4.8. *The Trace Scheduling Algorithm (Algorithm 4.2) terminates after a finite number of steps.*

PROOF. Assume that a sequential program G_{seq} has been given to Algorithm 4.2. Let N denote the total number of tests in the program. Then the total number of distinct paths through G_{seq} from START to END cannot exceed 2^N.

Take any operation A in the program. Let P denote the number of paths that contain A. Then $P \leq 2^N$. Suppose a trace $\mathcal{T}$ containing A is picked for the first time by Algorithm 4.2. The trace will lie on some of the paths through A. When $\mathcal{T}$ is replaced by its schedule using Algorithm 4.1, operation A may get knocked off some of the other paths that now contain it. To compensate for that, copies of A are inserted on those paths. A single copy typically serves a number of paths. Assume that k copies of A have been inserted: $A_1, A_2, \ldots, A_k$. Assume they lie on $P_1, P_2, \ldots, P_k$ paths, respectively. We then have

$$k \leq P_1 + P_2 + \cdots + P_k \leq P - 1, \tag{4.1}$$

since there must be at least one path through A that does not need a copy.

Next, suppose that at some point a trace containing A_1 is picked by Algorithm 4.2. As before, A_1 may drop out of some of the P_1 paths that contain it, and we may need to create several copies of A_1. Let $A_{11}, A_{12}, \ldots, A_{1\ell}$ denote these copies, and let them lie on $P_{11}, P_{12}, \ldots, P_{1\ell}$ paths, respectively. We then have

$$\ell \leq P_{11} + P_{12} + \cdots + P_{1\ell} \leq P_1 - 1, \tag{4.2}$$

since at least one path through A_1 does not need a copy. Since each of the $k - 1$ integers $P_2, \ldots, P_k$ is at least 1, it follows that

$$k - 1 \leq P_2 + \cdots + P_k. \tag{4.3}$$

Combining 4.1, 4.2 and 4.3, we get

$$\begin{aligned} \ell + (k-1) &\leq \ell + P_2 + \cdots + P_k \\ &\leq (P_1 - 1) + P_2 + \cdots + P_k \\ &= (P_1 + P_2 + \cdots + P_k) - 1 \\ &\leq P - 2. \end{aligned}$$

Note that when A is copied for the first time, the number of its copies available for future use is $k \leq P - 1$. When A_1 is copied, the number of copies of A available for future use is $(\ell + k - 1) \leq P - 2$.

The pattern is now clear: Every time a trace containing a copy of A is picked by Algorithm 4.2, the upper bound on the number of copies of A available for future use, decreases by at least 1. As the algorithm keeps running, eventually the number of available copies of A will become zero. This is true for each operation in the program. Since there are only finitely many operations to start with, after a finite number of steps there will be no operations or copies left to form a trace.[2] ■

[2]For an alternative approach to the termination proof, see [Nic84].

4.6 Picking Traces

We have not yet discussed which trace is selected for scheduling in Step 2 of Algorithm 4.2. The choice is important, as the order in which traces are chosen has a large impact on the parallelism that can be extracted from those paths—generally the traces selected first will be the longest and have the best schedules. The goal, then, is to pick the most important traces to schedule first. In the case of loop-free codes we want to select, at each step, the remaining trace that has the highest probability of being executed when the program is run. (Since the basic trace scheduling algorithm does not extend across loop boundaries, for programs with loops the strategy is to pick the most important traces within the most frequently executed loops.)

We associate a probability with each program node/edge representing the chance that it will be executed/selected during execution. As a first step, we associate probabilities with the branches of each test indicating the chance that a branch is taken given that control has reached the test. One standard approach is to use an execution profiler to record which way the branches go when a given program is run with (hopefully) representative inputs. Branch probabilities are then computed using this record.

To pick good traces it is very useful to know which way a given branch is likely to go, but very accurate estimates of branch probabilities are usually unnecessary (we elaborate on this point below). The following simplistic approach, which does not require a profiler, still often gives good results. For each edge e in the control flow graph connecting a node u to a node v, let $p(e)$ be the probability that e is taken when node u is reached. If u is not a test, the v is the sole successor of u and we set $p(e) = 1$. If u is a test, set $p(e) = .5$. The default that a test takes either branch with probability $.5$ can sometimes be improved with additional heuristics. For example, if it is clear that e is taken only when an error condition arises (e.g., because all paths from e end in jumps to exception handlers) then we may wish to set $p(e) = .01$ and $p(e') = .99$ for the other branch e'.

Given an estimate of the branch probabilities $p(\cdot)$ we can compute the execution and selection probabilities of $C : V \cup E \rightarrow [0, 1]$ for each node and edge. For each operation $v \in V$, $C(v)$ is the probability that v is executed, and for each control edge $e \in E$, $C(e)$ is the probability that e is selected. With S as the start node of the control flow graph, we compute $C(\cdot)$ beginning at S and proceeding with any topological order of the nodes and edges (i.e., $C(v)$ is computed after $C(e)$ is computed for all edges e that target v; similarly, $C(e)$ is computed after $C(u)$ is computed for the source node u of e):

$$
\begin{aligned}
C(S) &\leftarrow 1 \\
C(e) &\leftarrow C(u) \cdot p(e) && \text{for each edge } e : u \to v \\
C(v) &\leftarrow \sum_{e \in \text{in}(v)} C(e) && \text{for each node } v \in V - \{S\}
\end{aligned}
$$

To select a trace for scheduling, consider the set of operations not yet scheduled and select an operation v (the *seed*) with the highest value of $C(v)$. If there is a tie, break it arbitrarily.

Starting with the seed, first grow the trace forward in the direction of execution through the control flow graph. To do this, consider the set of successors to the node currently at the end of the trace and select one according to some criteria (see below). Add that successor to the end of the trace and repeat. Stop growing the trace forward when

- there is no successor to the node currently at the end of the trace,
- no successor satisfies the heuristic for choosing a successor,
- the successor chosen has already been scheduled, or
- the length of the trace has reached a pre-determined maximum.

Next, grow the trace backward from the seed. The process is the same, but we choose predecessors rather than successors to extend the trace.

The heuristic for selecting successors/predecessors to extend the trace from a node u usually depends on the execution probabilities of the edges leaving from or coming to u, and those of the successors and predecessors of u. An obvious choice of heuristic is to grow the trace forward by picking an edge $u \overset{e}{\to} v$ that has the highest value of $C(e)$ among all edges leaving u. To grow the trace backward, pick a predecessor w of u that has the highest value of $C(w)$ among all predecessors of u.

A different heuristic was used in the Bulldog [Ell85] and Multiflow compilers [LFK$^+$93]. To grow the trace forward, pick an edge

$u \xrightarrow{e} v$ such that $C(e)$ has the highest value among all edges leaving u, and also among all edges entering v. To grow the trace backward, pick an edge $w \xrightarrow{f} u$ such that $C(f)$ has the highest value among all edges entering u, and also among all edges leaving v. This heuristic has the disadvantage that sometimes no edge satisfies both criteria (while going forward or backward) and the trace stops growing, perhaps prematurely.

As a practical matter, it can be desirable for a trace scheduling compiler to either terminate or switch to less aggressive scheduling strategies after the most important traces have been scheduled. For example, to limit code growth, after a limit on compensation code has been reached, the compiler may decide to restrict traces to basic blocks to avoid the creation of more compensation code. In addition, at some point the probability of execution of the remaining traces is so low and the expected performance benefit of scheduling those traces so small that the compiler may simply leave the remaining sequential part of the program sequential. By adjusting the cut-off points for compensation code and expected benefit from scheduling traces, trade-offs can be made between code quality, compilation time and size of the resulting program.

Note that all of these decisions are heuristically made at compile-time and are dependent on compile-time estimates of the probability of execution of various paths, which are in turn dependent on the estimates of branch probabilities. And while it is true that even simple static estimates of branch probabilities are usually sufficient, this is a statement about the average or typical case, and there are many real and important scenarios where imprecise branch probabilities can and do lead to poor results. For example, if the compiler estimates that two paths are executed with probability .5 each, but the actual execution percentages are even moderately skewed, say .6 for path A and .4 for path B, then it is not unusual for the generated code to be suboptimal, and in pathological cases even worse than the original sequential program. This outcome can occur because decisions made in scheduling one trace may be at the expense of decisions made in scheduling a different trace. For example, if the compiler selects a trace including path B to schedule first, it may move operations above a jump on to the more frequently executed path A. These new operations may consume critical resources for many cycles relative to the length of A, causing the performance on path A to be worse than in the original sequential program. Furthermore, if there is not much parallelism on path B to begin with, the benefit of parallelizing B may not make up for the loss of performance on path A.

It is also worth mentioning that branch probabilities, even if accurate, are an abstraction and oversimplify the reality of program executions in at least two important ways. First, there is an assumption that different branches are not correlated—that if one test branches a particular way then that does not affect the probability of what branches other tests will take. In practice there are even perfectly correlated tests in real programs—consider the situation where the same flag is tested at the beginning and end of a function to handle some special case. Second, there is an assumption that the branch selected by a test is uncorrelated with the branch selected by the same test the next time it is executed. In practice some programs go through "modes" where certain paths are frequently executed for a period of time and then the mode changes and a different set of paths become most important. Despite these limitations, practical trace scheduling compilers have to date always used branch probabilities because they are easy to implement and work well in the great majority cases.

We conclude with a few comments on trace scheduling loops. To generate long traces for scheduling, the loop body can first be *unrolled* (see Section 6.2) some number of times and the resulting loop body is then trace scheduled. In fact, this is the approach adopted by all previous trace scheduling compilers. In principle, one could pick a trace that spans code preceding and following a loop, as well as the loop body itself. Such traces, however, are usually not chosen as they almost always contradict the fundamental heuristic on which trace scheduling is built, namely that traces are chosen in order of most likely to execute. Under this fundamental assumption, the code from before and/or after a trace would only be part of the same trace as the loop body only if the loop most frequently executed only a single iteration before terminating.

FURTHER READING

In [NF81], Nicolau and Fisher measure the upper bound on potential parallelism assuming an oracle to resolve conditional jump directions and indirect memory references during compaction. In [Nic84], Nicolau and Fisher study the available ILP for VLIW architectures. A preliminary evaluation of trace scheduling for global microcode compaction is presented in [GS83]. In [Lin83], Linn presents a generalization of trace scheduling for augmenting the use of global context information. In [SDX84], Su et al. propose heuristics for list scheduling of traces. Experiments over a limited set of test cases indicate that these heuristics facilitate code motion. Correctness of and disambiguation of flow-analysis in trace scheduling is discussed in [Nic84].

Compilation techniques for the early VLIW machine ELI (Enormously Long Instructions) are discussed in [FERN84]. In [CH88], Chang and Hwu study the different trace selection strategies and their relative effectiveness. In [JRS97], Jacobson et al. propose a path-based next trace prediction technique. Dynamic trace selection using hardware assists is discussed in [CHL+03]. Gross and Ward describe algorithms to remove redundancies in the compensation code [GW90]. A detailed discussion of techniques for avoidance and suppression of compensation code is available in [FGL94]. Techniques for predicting branch directions during trace scheduling are discussed by Fisher and Freudenberger in [FF92]. Bernstein et al. [BCK91] propose the *code duplication method* where operations are duplicated in order to schedule them earlier, thus providing a framework that facilitates moving of operations non-incrementally. In [ME92] Moon and Ebcioğlu presented a resource-constrained global scheduled technique for VLIW and superscalar processors. In [LFK+93], Lowney et al. provide a comprehensive description of the Multiflow trace compiler.

Although trace scheduling represents a major step forward towards global scheduling of sequences of basic blocks, it has been criticized for its narrow view, i.e., restricted to the current trace, of the program. Linn [Lin88] and Hsu [Hsu86] propose profile-driven algorithms for scheduling trees of basic blocks. An enhanced version of trace scheduling, called *Trace Scheduling-2*, is discussed in [Fis93]. A speculative approach for trace scheduling in VLIW processors is presented in [ANvEB02]. The success of trace scheduling led to the development of micro-architectural enhancements such as *trace cache* [RBS96, RBS99]. A discussion of trace processors is available in [RJSS97, RS99].

5

PERCOLATION SCHEDULING

Trace scheduling suffers from a number of problems related to its focus on a single trace at a time. Percolation scheduling overcomes these problems, to the extent possible at compile time, by providing a small set of transformations that can "percolate" operations upwards in a control flow graph so that they can be scheduled in parallel with other operations. These core transformations define the allowable motions of operations between adjacent nodes in the flow graph of a program at the operation level, where each node contains one or more operations that can be executed in parallel.

5.1 Introduction

In the preceding chapter, we presented trace scheduling as a global compaction technique to exploit parallelism beyond basic blocks. In trace scheduling, a trace is selected from the original (sequential) code and compacted. Compensation code is inserted at jumps into and out of the trace to ensure semantic correctness. Then a disjoint trace is selected and the process is repeated. Trace scheduling has the following limitations:

- Relatively localized detection and exploitation of parallelism, resulting from the finality and mutual exclusiveness of traces, i.e., traces are individually compacted and once a trace is compacted no operations can be moved into or out of that trace.

A. Aiken et al., *Instruction Level Parallelism*,
DOI 10.1007/978-1-4899-7797-7_5

- Trace scheduling can suffer from code explosion because it produces many redundant copies of operations. Although some copying is an inevitable effect of aggressive speculation, part of the copying in trace scheduling is a result of the granularity of the transformation itself, and part of it is due to the fact that the traces are treated separately. For instance, the fact that operations cannot cross trace boundaries implies that there is no way to combine copies of an operation from different paths.
- The effectiveness of trace scheduling is critically dependent on the predictability of conditional jumps at compile time and a heavy bias of the conditionals on the most important traces towards one of their branches.

Percolation scheduling (PS) [Nic85c] overcomes, to the extent possible at compile time, the problems that limit the effectiveness and applicability of trace scheduling. The goal of PS is to maximize parallelism by "percolating" operations upwards in a control flow graph so that they can be scheduled in parallel with other operations and as early as possible subject to dependences (and, in some versions of percolation scheduling, resource constraints). PS globally rearranges the code by moving operations from a node to a predecessor node (in a control flow graph) using a set of *core transformations*. The core transformations define the allowable motions of operations between adjacent nodes in the flow graph of a program at the operation level, where each node contains one or more operations that can be executed in parallel.

The core transformations are independent of any heuristics for prioritizing which operations to move and where to move them. The separation of the core transformations from superimposed heuristics (such as trace picking) facilitates exploitation of parallelism even in the presence of very unpredictable control flow.

5.2 The Core Transformations

In contrast to trace scheduling, the core transformations of percolation scheduling are defined locally, between adjacent nodes of a control flow graph. Furthermore, the core transformations can be used by a variety of higher-level strategies that direct which core transformations to perform. The only restriction placed on code motion by these transformations is that of respecting data dependences, which preserves the execution semantics of the original program. We first define the control flow graph of a program on which the core transformations of PS operate. Note that this control flow graph model is similar, but not identical, to the trace scheduling model, with the main difference being the more complex and general handling of branches in the PS model.

Let $\mathcal{N}$ denote the set of nodes $n_0, n_1, \ldots$ in a flow graph; node n_0 is the start node. We use two set-valued functions Pred and Succ, such that for each $n \in \mathcal{N}$, $\mathrm{Pred}(n)$ is the set of all immediate predecessors of n and $\mathrm{Succ}(n)$ is the set of all immediate successors of n. At any node, we denote the incoming edges by I_j and exiting edges by E_j.

Execution of a program begins at the start node and proceeds sequentially from node to node. When control reaches a particular node, all operations in the node are evaluated concurrently; the assignments update registers or memory locations and the conditionals return the next node in the execution sequence. Operations evaluated in parallel perform all reads before any assignment performs a write. Write conflicts within a node are not permitted.

A node may contain at most one conditional initially; however, as transformations are performed a node may come to contain multiple conditionals. PS models the set of conditionals in a node as a directed acyclic graph (dag) [KN85]. Each conditional in the dag has two successors corresponding to its true and false branches. A successor of a conditional is either another conditional or a pointer to a node. The dag of conditionals is (by construction) *rooted*, i.e., only one conditional (the root) has no predecessors. To evaluate a dag in a node in a given machine state, the (unique) path from the root to a leaf node is selected such that the branches on the path p correspond to the value (true or false) of the corresponding conditionals on p in the current state. Evaluation of the dag returns the node that terminates p. In practice (and in actual implementation [Ebc88]), the dag is reduced to an equivalent tree by splitting/duplicating (if necessary) shared subgraphs.

Given an arbitrary conditional in a dag (tree) of conditionals, we define three graph-valued functions: $s_p(x)$ denotes graph of operations above the conditional x, $s_t(x)$ denotes the graph of operations on the true branch of x, and $s_f(x)$ denotes the graph of operations on the false branch of x. For example, consider the node shown in

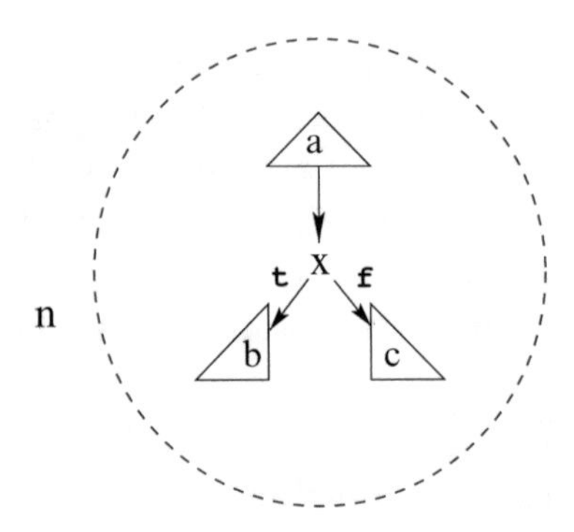

Figure 5.1 : A dag of conditionals

Figure 5.1. Node n consists of a dag of conditionals, where x is a conditional operation and a, b, c are dags of conditionals themselves. In this case, $s_p(x) = a,\ s_t(x) = b,$ and $s_f(x) = c$.

5.2.1 Delete Transformation

The *delete* transformation removes a node from the flow graph if it is empty (contains no operations) or unreachable. A node may become empty or unreachable as a result of other transformations. For example, in Figure 5.2 the empty node n (represented by a dashed circle) is deleted from the flow graph. Note that an empty node has exactly one successor. The transformation is defined as follows:

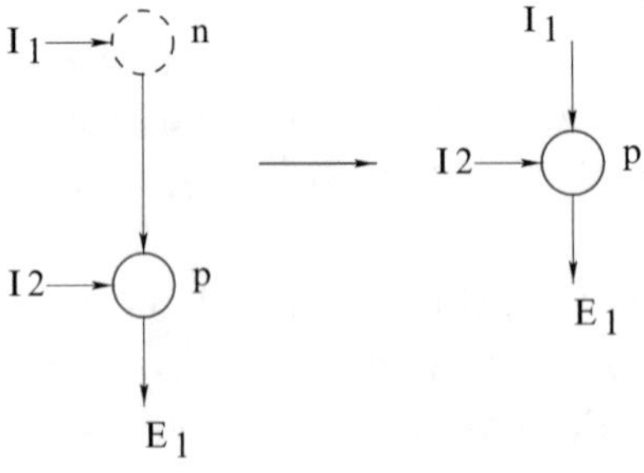

Figure 5.2 : Delete Transformation

Transformation Delete (m, n).

for each empty/unreachable node n **do**
  for each node $m \in \mathrm{Pred}(n)$ **do**
    $\mathrm{Succ}(m) \leftarrow \mathrm{Succ}(n)$
  endfor
endfor

5.2.2 Move-op Transformation

The *move-op* transformation moves an assignment operation x from a node n to a node m along the edge $\langle m, n \rangle$ provided no conflict exists between the operations in x and the operations in m and x does not kill any live value at m. Care must be taken not to affect the computation along the paths passing through n but not through m. To ensure this, the original node n is copied along all such paths (which can all share the same node n', rather than creating multiple copies along each path). For example, in Figure 5.3 the assignment operation x is moved from node n to node m. In order to preserve program semantics, node n is duplicated (n') along the path corresponding to the incoming edge I_2. The transformation is defined as follows:

Transformation Move-op (x, m, n). The transformation assumes that there does not exist dependences between any operation in node m and the operation being moved from node n along the edge $\langle m, n \rangle$ and that x does not kill any live values on other edges emanating from m, besides $\langle m, n \rangle$, if any.

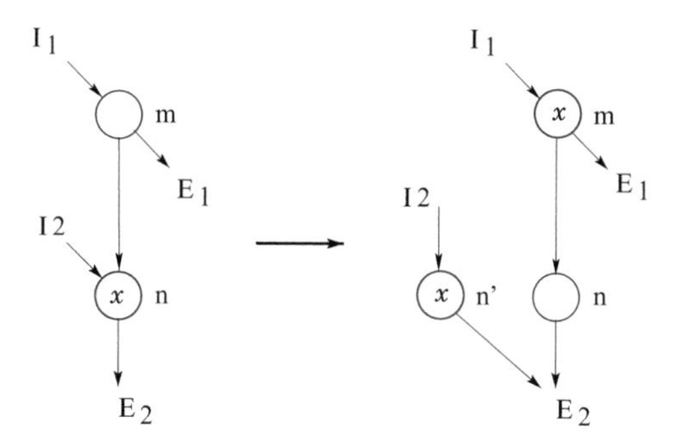

Figure 5.3 : Move-Op Transformation

// Duplicate n
for each node $p \in \text{Pred}(n) - \{m\}$ **do**
  Create a copy n' of node n
  $\text{Pred}(n') \leftarrow p$
  $\text{Succ}(n') \leftarrow \text{Succ}(n)$
endfor
Add x to node m
Delete x from node n

5.2.3 Move-test Transformation

The *move-test* transformation moves a conditional x from a node n to a node m through an edge $\langle m, n \rangle$ provided that no dependence exists between x and the operations of m; it also creates a copy of the original node n along all the paths passing through n but not through m to preserve program semantics. The test being moved may come from an arbitrary point in a dag of conditionals. Thus, node n itself is "split" into two nodes n_t and n_f, where n_t and n_f correspond to the true and false branches of x. As an illustration, consider the example shown in Figure 5.4, where a represents a dag of conditionals (in n) preceding x ($= s_p(x)$), b represents a dag of conditionals reached on x's true branch ($= s_t(x)$), and c represents a dag of conditionals reached on x's false branch ($= s_f(x)$). The non-control operations of n are *op*(n). The transformation moves the conditional x to node m and splits the node n into two, corresponding to the true and false

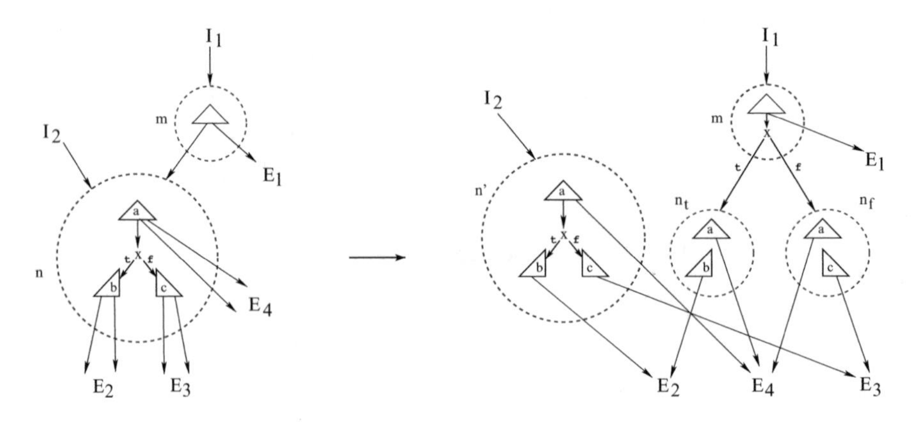

Figure 5.4 : Move-test Transformation

branches of x. The node n is duplicated (n') along the path corresponding to the incoming edge I_2.

Transformation Move-test (x, m, n). The transformation assumes there does not exist any dependence between any operation in node m and the conditional being moved from node n along the edge $\langle m, n \rangle$. Let the false branch edge of a conditional x be x_f and the true branch edge be x_t.

// Duplicate n
for each node $p \in \text{Pred}(n) - \{m\}$ **do**
  Create a copy n' of node n
  $\text{Pred}(n') \leftarrow p$
  $\text{Succ}(n') \leftarrow \text{Succ}(n)$
endfor
// Create nodes n_t, n_f corresponding to the true and false branches of x
// such that they contain all non-control operations of n
$n_t \leftarrow s_p(x) \cup s_t(x) \cup op(n)$
$n_f \leftarrow s_p(x) \cup s_f(x) \cup op(n)$
Move x to m, changing each edge in m pointing to n to point to x
Connect x's true branch to n_t, the false branch to n_f
Delete n

Note that creation of n_t entails explicit linking of $s_p(x)$ with $s_t(x)$. Likewise, the creation of n_f entails explicit linking of $s_p(x)$ with $s_f(x)$.

5.2.4 Unify Transformation

The *unify* transformation moves a copy of identical assignment operations x from a set of nodes n_j to a common predecessor node m. The transformation is legal if no dependence exists between x and the operations of m and x does not kill any value live at m on paths along which x is not moved. As in previous transformations, a node is copied along each path passing through the n_j but not through m to preserve program semantics.

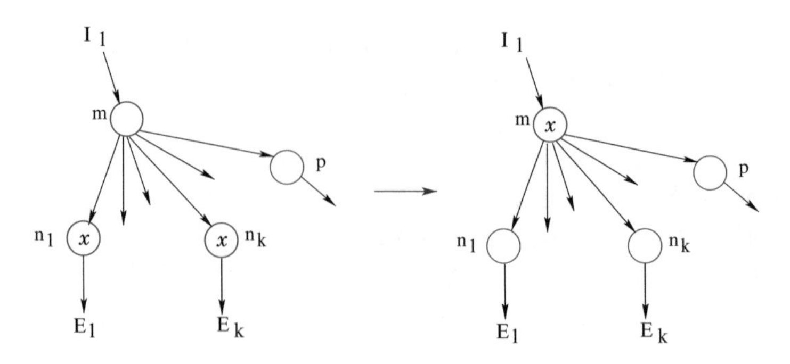

Figure 5.5 : Unify Transformation

Transformation Unify (x, m, n_j). The transformation assumes there does not exist a dependence between any operation in node m and the operation being moved from the set of nodes n_j along the the set of edges $\langle m, n_j \rangle$.

Add x to node m
for each node $p \in n_j$ **do**
    // Duplicate p
    Create a copy p' of node p
    for each node $q \in \text{Pred}(p) - \{m\}$ **do**
        $\text{Pred}(p') \leftarrow q$
        $\text{Succ}(p') \leftarrow \text{Succ}(p)$
    endfor
    Delete x from node p
endfor

Note that the core transformations of PS discussed above can be applied arbitrarily or can be applied in a meaningful order. For now, in the following example we show the results of applying the transformations in a good order; how to derive such an order is discussed later in this chapter.

Example 5.1. In this example we illustrate percolation scheduling and contrast it with trace scheduling. Consider the program and its flow graph as shown in Figure 5.6.

As discussed in Chapter 4, selected traces cannot extend past loop entrances and exits. Assuming equal probability for the branches of

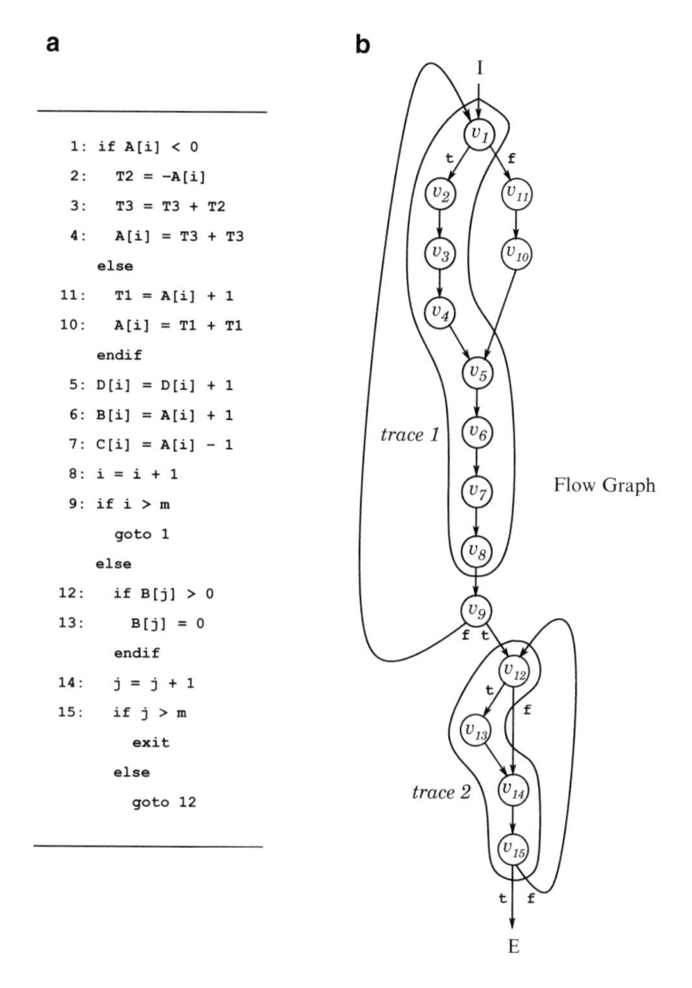

Figure 5.6 : A sample program and its flow graph

the inner conditionals, the traces $T_1 = \{v_1, \ldots, v_9\}$ and $T_2 = \{v_{12}, \ldots, v_{15}\}$ are picked. Subsequently, each trace is compacted and the corresponding compensation code is generated as explained in Chapter 4. The final schedule obtained from trace scheduling is shown in figure 5.7(a).

During percolation scheduling, the operations are moved to the top as far as possible, subject to the dependences, using the core transformations. For example, operation v_5 is copied on both the branches of operation v_1 and then the copies are unified at the top using the *unify* transformation. The final schedule obtained from percolation

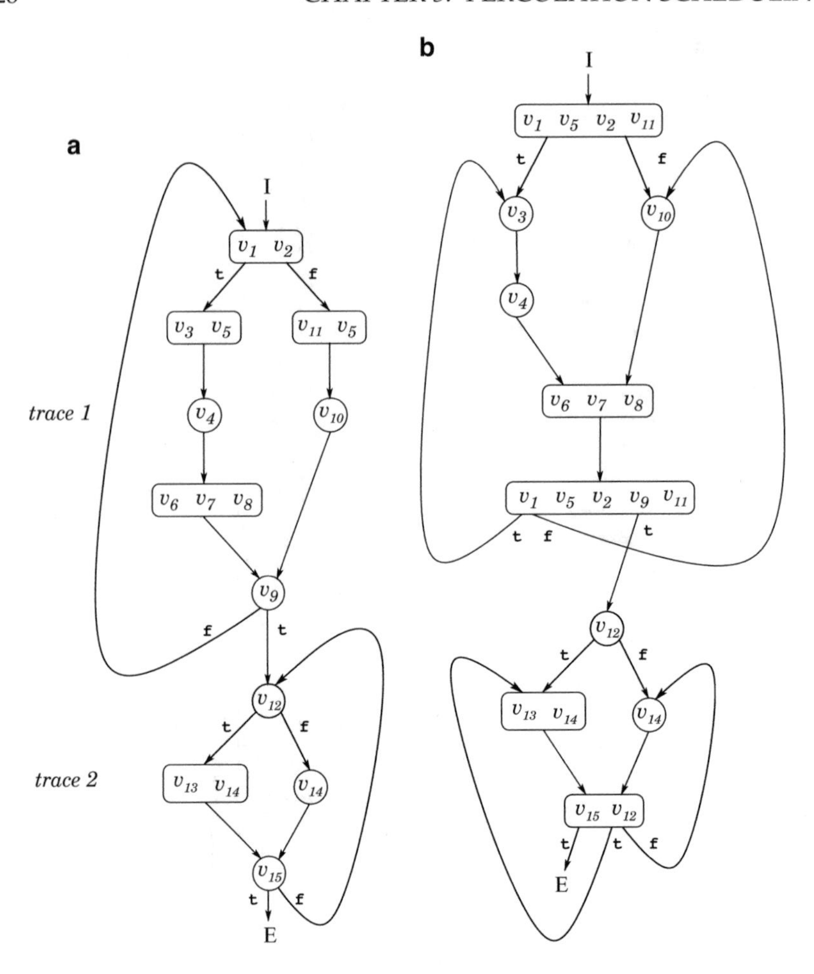

Figure 5.7 : Trace Scheduling vs. Percolation Scheduling

scheduling is shown in figure 5.7(b). Note that percolation scheduling of the loops results in a shorter (faster) schedule than the schedule produced by trace scheduling.

5.3 Remarks

Mechanisms such as disambiguation, renaming [KKP+81], loop quantization [Nic85a, Nic88] and other traditional optimizations can potentially play a vital role in exploiting parallelism in conjunction with the core transformations. However, these are not themselves scheduling transformations.

A complication that we have not mentioned thus far is the treatment of operations that can raise exceptions. Any technique that rearranges the order of branches and other operations potentially can (incorrectly) alter the meaning of a program by moving an operation (e.g., a division) above a conditional protecting that operation from raising an exception (e.g., a check to see if the divisor is zero). Note that such transformations do not violate any explicit data dependences, because both the test and the operation only read the value in question and the exception side-effect is implicit.

There are two ways to ensure correct exception behavior with percolation scheduling (or, again, any other technique that moves operations above conditionals). One is to simply not move operations that can raise exceptions above conditionals or, more generally, not move them above any conditional that may change whether or not the operation can raise an exception. If the hardware supports *deferred exceptions*, then when an operation faults it produces only a distinguished value that indicates an exception must be raised if the value of the operation is actually needed by the computation. On such an architecture, an operation x can be moved above conditionals by effectively splitting x in two: the operation proper, which may produce a deferred exception, and a test (that cannot be moved) left in the place of x that raises a real exception if the result of the operation was a deferred exception. Thus, even if the operation faults, no exception will be raised unless control actually comes on to the path where x would have necessarily been executed.

5.3.1 Termination

Code motion based on the transformations discussed in the previous section form the basis of percolation scheduling. It is important to guarantee that the process of applying the transformations terminates, i.e., given a flow graph, after a finite number of applications of the transformations, the transformations cannot be further applied to the resulting flow graph. Intuitively, termination of PS stems from the fact that the depth of a node, defined as the maximum number of nodes between that node and the start of any path reaching that node, decreases as a result of the application of the transformations. While the number of paths and the number of operations may increase during PS, the length of the paths does not. This guarantees termination as the upward motion of an operation is bound by its depth in the original flow graph; a detailed proof is available in [Nic86]. The termination argument applies to non-cyclic flow graphs. For loops, guaranteeing termination can be done either by limiting transformations across backedges or by using the algorithms presented in subsequent chapters.

5.3.2 Completeness

Just as the core percolation scheduling transformations can achieve some code motions that trace scheduling cannot, one might ask whether there is something more general than percolation scheduling —i.e., is percolation scheduling *complete* in the sense that it allows every possible legal code motion? We now present an informal outline of the argument for the completeness of the core transformations with respect to all the possible incremental (step-by-step, one operation at a time) parallelization code motions; a formal argument can be found in [Aik88].

The *move-op* and *unify* transformations cover all local (i.e., between adjacent nodes) code motions of operations, subject to dependences. Similarly, the *move-test* transformation covers all local code motions of conditionals. The *delete* transformation can be seen as a cleanup phase for other transformations; it does not change the relative order of operations in the schedule. Finally, backward-movement (away from the start node) transformations cannot achieve better results than the core transformations as moving one operation "down" is the same as moving all other operations "up". Hence, the core transformations are complete in the sense that if there is any legal motion of an operation to an earlier place in the schedule, constrained only by data dependences, then that code motion can be achieved by some sequence of the core transformations.

It is important to understand that the completeness result only holds with respect to local (node-to-node) transformations within an acyclic control-flow graph. There are non-incremental, non-local transformations for loops that percolation scheduling cannot express. One such non-local transformation is loop invariant code motion, which is an important optimization even for non-parallelizing compilers and can also be instrumental in facilitating exploitation of ILP beyond what is exposed by percolation scheduling.

5.3.3 Confluence

Confluence, also called the diamond property [Bar84], shows that the final result of a sequence of transformations is independent of the choice of transformation at each step in the sequence. This is a desirable property—if a system of transformations is confluent, then there is no concern that a sequence of transformations can lead to a local optimum from which it is impossible to proceed to a global optimum.

Unfortunately, the core transformations of percolation scheduling are not a confluent system [Aik88]; the non-confluence stems from the possible presence of false dependences in the flow graph. False dependences cause difficulties for many compiler optimizations, not just instruction scheduling, and various approaches have been proposed to deal with or eliminate false dependences [KKP+81, AKPW83, Wol78, CF87]. In absence of false dependences the set of core transformations form a confluent system.

5.4 Extensions

Although the repeated application of the core transformations can eventually move an operation as far as possible (towards the top) in the flow graph, the transformations themselves do not prescribe any order for applying the transformations. Given that the core transformations are not confluent, picking a good order in which to apply the transformations is clear important. In the rest of the subsection, we discuss techniques to guide the percolation process so as to improve the efficiency of PS.

5.4.1 Migrate Transformation

Migrate is guaranteed to percolate an operation as far as possible, subject only to data dependences. Migrate percolates the copies of an operation along with the operation itself, so as to enable as many copies as possible to be unified at common ancestors in the control flow graph. Intuitively, the transformation moves the copies of an operation that are deepest first, up to a point just below a common ancestor with other copies of the same operation. When as many copies as possible are in immediate successors of the common ancestor, unification(s) are attempted. The process of moving operations up followed by unification at common ancestors is repeated until no further motion is possible.

More precisely, each node in the graph is labeled with its depth using a modified depth-first search [Luc91, Tar95, Tar72, HT73]. When a node is duplicated, it inherits the label of the original node. While this may not be the actual depth of the copied node in the flow graph, it is a consistent labeling: all ancestors of the copied node have lesser labels and all successors have greater labels. The algorithm always picks an operation in a node with the largest label to move next, attempting unifications whenever possible. The reader is referred to [Aik88] for further details.

5.4.2 Trailblazing

Although PS provides an effective means of exploiting ILP the fact that it moves operations only between adjacent nodes causes two problems. First, even in the case where the operation to be moved is independent of every node along a path, each node on the path must be visited to move the operation from the end to the start of the path. Furthermore, the sequence of local moves often results in code duplication, even if it is legal to move the operation non-locally across all of the intermediate nodes.

Trailblazing [NN95] adopts a hierarchical approach to percolation scheduling. The program is represented hierarchically and some nodes represent compound statements in the original program. Moves are still local in the sense that transformations are between adjacent nodes, but a single transformation can now move an operation across many nodes in the original, flat control flow graph. Trailblazing uses a *hierarchical task graph (HTG)* [Gir91, GP92] and extends the core transformations to navigate the HTG hierarchy. An HTG consists of:

- *simple nodes* representing individual VLIW instructions, i.e., they contain one or more operations that can be executed in parallel, just as in the non-hierarchical version of percolation scheduling,

- *compound nodes* representing basic blocks, subroutines and if-then-else blocks, and

- *loop nodes* encapsulating loops.

The compound nodes act as "bridges" across which operations can be moved without visiting any of the nodes within, thus eliminating code explosion that may be caused by moving an operation using PS. Trailblazing attempts to move operations across these compound and loop nodes (past which PS or TS and its derivatives is unable to move operations at all). A detailed description of the algorithm is in [NN95].

5.4.3 Resource-Constrained Percolation Scheduling

Trace scheduling and PS generate idealized schedules that do not take resources into account. The compacted schedule is heuristically mapped on to a given machine (e.g., by using a greedy resource allocation algorithm). This is straightforward for each trace, though some additional patching up may be needed at trace exits/entry, and can also be easily adapted for PS. However, the drawback of this approach is that it is very hard to maintain a global view of the program while doing resource mapping, particularly across potential execution paths in the flow graph. Furthermore, the approach may have to undo some of the code motions performed during the compaction phase. This situation is complicated by the fact that some of the transformations do not have a unique inverse, and undoing them may potentially degrade performance.

Resource-constrained percolation scheduling (RCPS) separates the transformations from the resource mapping heuristics in the sense that first a set of operations that can be potentially scheduled (obtained after applying the PS transformations) is found and then the k most important operations are selected, where k is the maximal number of resources available. The algorithm consists of two phases: first, information about which operations can be moved and where is gathered, and then the operations are scheduled through PS transformations while taking resources into account. Thus, RCPS provides both a top-down view with respect to the "filling" of nodes guided by the resource constraints, and a bottom-up view, with respect to percolation of operations up the schedule, to the scheduling process. RCPS uses the *migrate* transformation to move operations to the targeted node and then maps the operations to the resources in a heuristically-defined order (since optimal scheduling with resource constraints is NP-Complete [GJ79]). A detailed description of the algorithm is presented in [EN89].

6

MODULO SCHEDULING

Loop parallelization, and particularly the parallelization of innermost loops, is the most critical aspect of any parallelizing compiler. Trace scheduling can be applied to loops, but has the disadvantage that loops must be unrolled to exploit parallelism between loop iterations, which can lead to code explosion and in general still does not exploit all of the parallelism available. In this chapter we discuss modulo scheduling, which was the first technique to address the scheduling of loops (both within and across iterations) directly and is still very widely used in practice. The original modulo scheduling technique applies only to loops where the loop body is a single basic block; we also present extensions to modulo scheduling that allow the technique to be applied to more general loops.

6.1 Introduction

Loop parallelization is the most critical aspect of any parallelizing compiler, since most of the time spent in executing a given program is spent in executing loops. In particular, the innermost loops of a program (i.e., the loops most deeply nested in the structure of the program) are typically the most heavily executed. It is therefore critical for any parallelizing compiler to try to expose and exploit as much parallelism as possible from these loops.

The simplest approach to extracting ILP from loops is to schedule the loop body. This method can find parallelism within a single loop iteration, but cannot exploit parallelism that may exist between different iterations of a loop. Furthermore, some loop bodies are just

A. Aiken et al., *Instruction Level Parallelism*,
DOI 10.1007/978-1-4899-7797-7_6

short; any instruction scheduling technique will struggle to extract parallelism if the loop body contains only a handful of instructions.

Historically, *loop unrolling*, a standard non-local transformation, is used to expose parallelism beyond iteration boundaries and expose a large number of instructions for scheduling. When a loop is unrolled, the loop body is replicated to create a new loop. Compaction of the unrolled loop body helps exploit parallelism across iterations as the operations in the unrolled loop body come from previously separate iterations. However, usually loops cannot be fully unrolled, either because the loop bounds are unknown at compile time or, very frequently, because high degrees of unrolling result in serious compiler performance problems due to space consumption. Of course, less unrolling also limits the amount of instruction-level parallelism exposed.

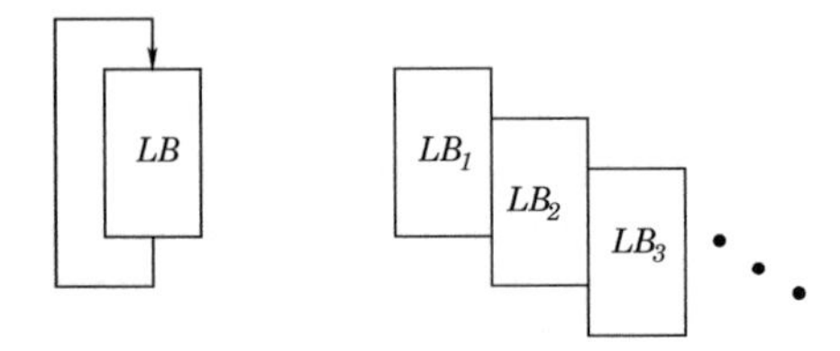

Figure 6.1 : Loop pipelining

In order to extract higher levels of parallelism in software without unrolling, the various iterations of a loop can be pipelined (akin to pipelining of operations in hardware [PD76, Kog77b, Kog81]) subject to dependences and resource constraints. This technique is known as *loop pipelining*. An example of loop pipelining is shown in Figure 6.1. Figure 6.1(b) shows the pipelined execution of the different iterations of the loop shown in Figure 6.1(a), where LB_i represents the loop body of the i-th iteration. Loop pipelining has its roots in the hand-coding practices for microcode compaction [TTTY78, Cha81, Tou84].

Like loop pipelining, *Modulo scheduling* [RG81, Hsu86] initiates successive iterations of a loop at a regular, fixed interval called the *initiation interval* or *II*. Unlike loop pipelining, modulo scheduling also schedules the loop body, including allowing the schedule for the loop body to be made *longer* by introducing holes in the schedule if that results in a smaller initiation interval. The combination of initiation interval and selected schedule for the loop body must respect the dependences of the original loop and the resource constraints of the target machine.

Modulo Scheduling was the first method proposed for *software pipelining* (the term software pipelining, originally coined to describe a specific transformation [Lam88], is nowadays commonly used interchangeably with loop pipelining to describe any and all transformations that deal with extracting instruction-level parallelism from loops). The objective of software pipelining is to find a recurring pattern, or *kernel*, across iterations, and to replace the original loop with this kernel, plus prologue and epilogue code. The problem of loop pipelining then reduces to how to determine the (best) parallelized kernel for a loop.

6.2 Unrolling

Before discussing modulo scheduling we first discuss loop unrolling, which is useful both to motivate the benefits of modulo scheduling but is also an important loop transformation that we will use in the subsequent development. As mentioned above, the earliest technique for extracting instruction level parallelism from within and across multiple iterations was *unrolling* or *unwinding* of the loop. If the number of iterations in the loop is a compile time constant, the loop can be *fully unrolled* by simply repeating the loop body a number of times equal to the number of iterations of the loop. For a `for`-style loop, we of course also need at least to insert the update of the iteration variable between the copied iterations, but in practice it is more common to modify the copies of the loop body to replace references to the loop index with an expression for the loop index's value in each loop iteration. For example, if the loop index is a variable i that ranges from 1 to 10 over 10 iterations of the loop, then occurrences of i in the first unrolled iteration would would be replaced by 1, occurrences of i in the second unrolled iteration would be replaced by 2, and so on. In either case the loop itself (i.e., the exit tests and jumps associated with the loop structure) is eliminated, which reduces the number of instructions executed at run-time and therefore potentially reduces the execution time even if no parallelism is extracted. Because of this property, full unrolling is sometimes used as an optimization even in architectures exhibiting no parallelism. An example of full unrolling is given in Figure 6.2.

```
do i = j, j+3                  a[j] = 0.0
     a[i] = 0.0                a[j+1] = 0.0
end do                         a[j+2] = 0.0
                               a[j+3] = 0.0
```

Figure 6.2 : Full unrolling: before and after unrolling

Note that the loop bounds themselves need not be known at compile time for full unrolling to be feasible: all that is required is that the number of iterations be constant and known at compile time. Also note that this implies that `repeat` or `while`-style loops generally cannot be fully unrolled.

Of course, full unrolling is not always possible, since the number of iterations is usually not a compile-time constant. Furthermore, even in cases where the number of iterations is a compile-time constant, unrolling is impractical if the number of iterations is large. Even if space per se is not a concern and thus the code explosion resulting from full unrolling of loops could be tolerated (at least for innermost loops) such unrolling may actually be undesirable from a performance point of view. Indeed, the unrolled loop may adversely affect instruction cache performance, possibly degrading overall performance despite the reduction in the number of instructions executed.

For these reasons, full unrolling is usually not done. However, a partial unrolling is often useful, and in fact partial unrolling can always be performed in both `for`-style loops (where the termination condition is independent of what is computed in the loop body, such as in Figure 6.2) and in `while`-style loops (where the termination condition depends on what is computed on each iteration, for example checking for a string or list terminator). Partial unrolling can take two forms. The simplest and most general approach consists of unrolling the loop by replicating the loop body, including the exit test, a fixed number of times (called the *unrolling factor*), replacing the original loop body with this unrolled loop body. If the loop has a step, the loop step is also changed by multiplying the original step by the unrolling factor. An example is given in Figure 6.3.

```
                    do i = 1, N, 2
                        a[i] = 0.0
                    end do

    After unrolling:
                    do i = 1, N, 6
                        a[i] = 0.0
                        if i >= N exit
                        a[i+2] = 0.0
                        if i + 2 >= N exit
                        a[i+4] = 0.0
                    end do
```

Figure 6.3 : Unrolling a simple loop three times

If the compiler can determine that the number of iterations executed is a multiple of the unrolling factor,[1] we can perform a further optimization by removing the intermediate exit tests in the unrolled loop body [Ell85, CNO+88]. Even if the number of loop iterations is statically unknown, the transformation can still be done if the number of loop iterations is known at runtime when the loop begins executing. The original loop is replaced by two loops. The first loop (possibly unrolled with exit tests left in) executes just enough iterations so that the remaining number of iterations is evenly divisible by the desired unrolling factor. The second loop is then unrolled using the chosen factor and the intermediate exit tests are removed. For loops with a sufficiently large number of iterations, this transformation potentially results in substantial performance benefits. An example is given in Figure 6.4. Due to the overhead involved this method is not used as often as one might expect. Also note that this method is not applicable to `while`-style loops where the number of loop iterations is unknown even during execution of the loop (i.e., it is known only when executing the final test that determines the loop terminates, not prior to executing the loop).

Once the loop has been unrolled, any of the scheduling techniques discussed in previous chapters (e.g., list scheduling, trace scheduling or percolation scheduling) can be applied to extract ILP both within

[1]In practice this means that we set the unrolling factor such that it is an even divisor of the number of iterations of the loop, if known.

```
                do i = 1, N
                  a[i] = 0.0
                end do

After test removal:
                do j = 1, N mod 3
                  a[j] = 0.0
                end do
                do i = j, N, 3
                  /* Assume j is available (after increment)
                     after first loop completes */
                    a[i] = 0.0
                    a[i+1] = 0.0
                    a[i+2] = 0.0
                end do
```

Figure 6.4 : Test removal when number of iterations is only known at runtime

and across iterations of the original loop. A limitation of this approach is that parallelism cannot be extracted across iterations of the unrolled loop. For example, say a loop L is unrolled three times to create a loop L^3 (e.g., as in Figure 6.4). The scheduling techniques we have discussed so far can parallelize the loop body of L^3, effectively overlapping iterations 1–3 and 4–6 of L, but no parallelism can be exploited between, say, iterations 3 and 4 of L because they are in different iterations of L^3. In the rest of this chapter and subsequent chapters we discuss techniques that can extract parallelism across all iterations of a loop.

6.3 Preliminaries

A data dependence between operations from the same iteration is a *intra-iteration dependence*; a data dependence between operations from different iterations is a *inter-iteration* or *loop carried* dependence. For example, in Figure 6.5, the dependence between operation v_2 and operation v_1 is an intra-iteration dependence; whereas the dependence between v_2 and v_3, shown by a dashed arrow in the data dependence graph (DDG) shown in Figure 6.5 b), is a loop carried dependence. The dependence *distance* (in number of iterations) between v_2 and v_3 is shown in angled brackets.

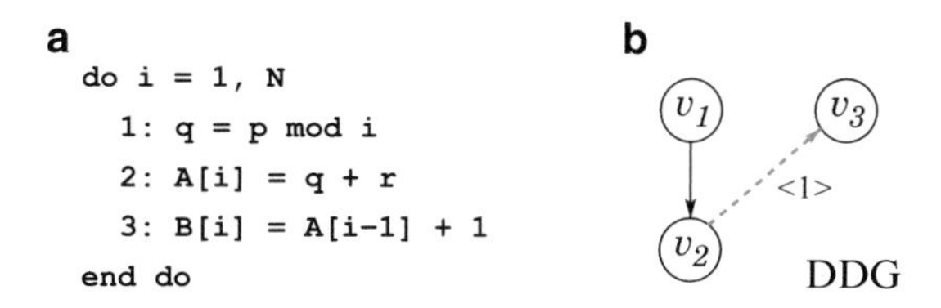

Figure 6.5 : Intra-iteration and loop carried dependences

Definition 6.1. *Given a dependence graph $G(V, E)$, a path from an operation v_1 to an operation v_k is a sequence of operations $\langle v_1, \ldots, v_k \rangle$, such that $(v_{i-1}, v_i) \in E$ for $i = 1, 2, \ldots, k$.*

Definition 6.2. *A* cycle *in a dependence graph is a path $\langle v_1, \ldots, v_k \rangle$ such that $v_1 = v_k$ and the path contains at least one edge. A cycle is* simple *if, in addition, $v_1, \ldots, v_{k-1}$ are distinct.*

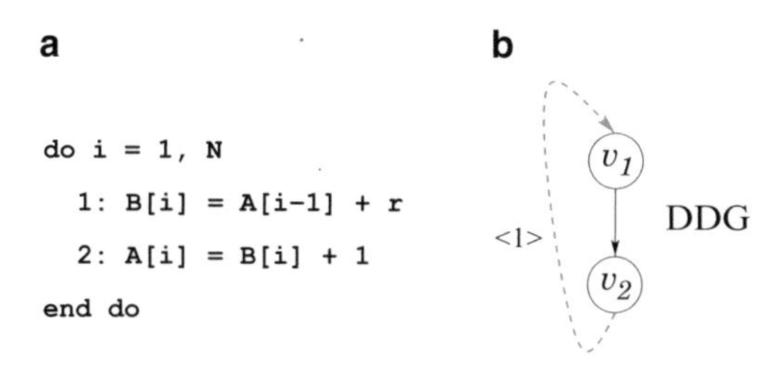

Figure 6.6 : Loop recurrences

A loop carried dependence in conjunction with intra-iteration dependences may form a simple cycle in the dependence graph. For example, in Figure 6.6, the intra-iteration dependence and the loop carried dependence between the operations v_1 and v_2 form a simple cycle. A loop has a *recurrence* if an operation in one iteration of the loop has a direct or indirect dependence upon the same operation from a previous iteration, e.g., in Figure 6.6 operations v_i^k and v_i^{k-1} constitute a recurrence, where v_i^k represents the i-th operation of the k-th iteration. In general, a recurrence may span several iterations, i.e., an operation v_i^k may depend on an operation v_i^{k-j}, where $j \geq 1$. The existence of a recurrence manifests itself as a simple cycle in the dependence graph. Subsequently, a cycle refers to a simple cycle in a dependence graph.

The *length* of a cycle c, denoted by $len(c)$, is the sum of the latencies of operations in c. The *delay* of a cycle c, denoted by $del(c)$, is the

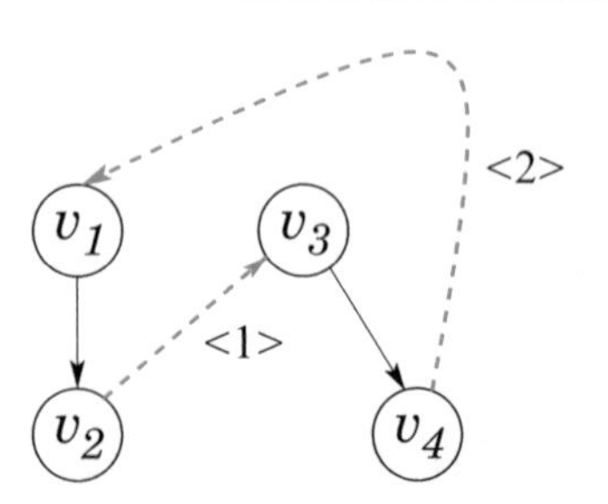

Figure 6.7 : A simple cycle

time interval between the start of operation v_1 and the start of operation v_{k-1} in a given schedule plus one, where operations $v_1, v_{k-1} \in c$, see Definition 6.2. Let $dist(c)$ denote the sum of the distances (corresponding to all the loop carried dependences) along c. For example, in Figure 6.7 the path $\langle v_1, v_2, v_3, v_4, v_1 \rangle$ constitutes a simple cycle. The length of the cycle is 4 and $dist(c) = 1 + 2 = 3$.

Note that the delay of a cycle c may not be equal to its length as the delay of a cycle is dependent on the actual scheduling of operations constituting the cycle. In general, for any cycle c in a dependence graph, $len(c) \leq del(c)$.

Finally, consider two operations o_1 and o_2 where o_2 comes after o_1 in the program and o_2 also depends on o_1. Recall that there are three distinct kinds of dependence that could exist between o_1 and o_2 (Definition 3.1): A true (or data) dependence when o_2 reads a location that o_1 writes, an anti-dependence when o_2 writes a location that o_1 reads, and an an output dependence when o_1 and o_2 write the same memory location. Anti- and output dependences can set up recurrences in the data dependence graph which potentially degrade the ability to extract ILP. To avoid this problem, the *dynamic single assignment form* [Fea88, Fea91, Rau92] may be used to minimize the anti- and output dependences. We will focus on true dependences in our examples, but the techniques we present apply to all dependences, regardless of type.

6.4 Modulo Scheduling Algorithm

Modulo scheduling takes as input a loop with a single basic block for the loop body and searches for a combination of an initiation interval and a schedule for the loop body that satisfies both the machine's resource constraints and the data dependences of the loop. The essential idea is to fix an initiation interval i and then apply a basic block scheduling algorithm (e.g., list scheduling) to the loop body, with two modifications. First, when the initiation interval is i and operation o is scheduled, that means that an instance of operation o will be issued every i control steps. Thus, we record the fact that the resources required by o are in use every ith step. Second, the basic block scheduler only considers the intra-iteration dependences. When a schedule for the loop body is found we must still verify that the loop carried dependences are satisfied. If modulo scheduling encounters a situation where an operation cannot be scheduled due to resource constraints or a schedule for the loop body does not satisfy loop carried dependences, the initiation interval is increased by 1 and the entire process is repeated. The algorithm is guaranteed to terminate, because when the initiation interval reaches the length of the original loop body, modulo scheduling degenerates to scheduling only the loop body and not overlapping iterations, which will always succeed.

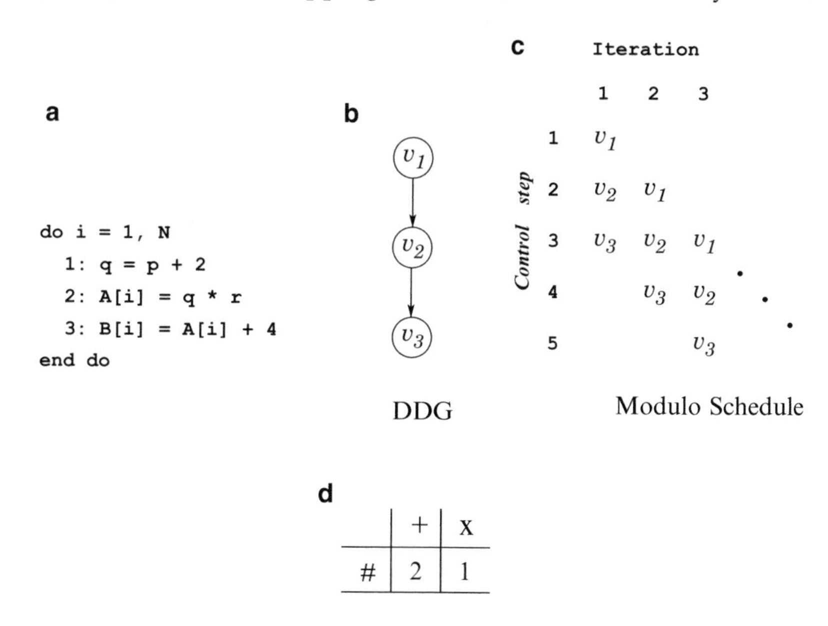

	+	x
#	2	1

Figure 6.8 : An overview of modulo scheduling

An illustration of modulo scheduling is shown in Figure 6.8. Consider the loop and its data dependence graph shown in Figures 6.8(a) and 6.8(b) respectively. For the resources given in Figure 6.8(d) and a single cycle latency for each resource, a modulo schedule is shown in

Figure 6.8(c). The reader may find it instructive to verify the schedule with respect to the dependences and resource usage.

Let $\mathcal{R}$ be the resources available for scheduling loop L. Resources are grouped into classes, each of which consists of identical resources. Let $\mathcal{R}(i)$ denote the i-th resource class, and let the number of resources in the i-th class be denoted by $num(i)$. For example, in Figure 6.8(d), $\mathcal{R}(1)$ represents the set of adders, where $num(1) = 2$. Similarly, $\mathcal{R}(2)$ represents the set of multipliers, where $num(2) = 1$. For simplicity of exposition, each resource is assumed to have single cycle latency, but the techniques discussed here extend easily to deal with multi-cycle resources [Rau95].

Let V denote the set of operations in L's loop body. The function $r : V \rightarrow \{1, 2, \ldots\}$ gives the number of the resource class $r(v)$ used by each $v \in V$. (For simplicity, we assume each operation requires only one resource.) For example, in Figure 6.8(d), $r(v_1) = 1$ and $r(v_2) = 2$.

Algorithm 6.1. Modulo Scheduling
The input to this algorithm is a dependence graph $G(V, E)$ of the body a loop L. The output is an initiation interval ii and a modulo schedule assigning a control step $l(v)$ to each $v \in V$. An iteration is initiated every ii cycles and executes the schedule for the loop body given by the $l(v)$. Pseudo-code for modulo scheduling is given in Figure 6.9.

The algorithm given in Figure 6.9 can be improved in a number of ways. It is rarely necessary to start with an initiation interval of 1, because it is usually possible to quickly compute a better lower bound on the smallest possible initiation interval. Modulo scheduling implementations first look at the loop body and compute the *resource-constrained initiation interval (ResII)*, a lower bound on the initiation interval due to resource constraints:

$$\text{ResII} = \left\lceil \max_i \frac{|\{v|v \in V \text{ and } r(v) = i\}|}{num(i)} \right\rceil \tag{6.1}$$

Next, the *recurrence-constrained initiation interval (RecII)*, a lower on the initiation interval based on the length of recurrences in the dependence graph, is computed:

$$\text{RecII} = \left\lceil max_{c \in C} \frac{len(c)}{dist(c)} \right\rceil \tag{6.2}$$

where C is the set of all simple cycles in the dependence graph.[2]

[2] The recurrences in the dependence graph can be enumerated using the algorithms in [Tie70, MD76]. RecII can also be computed in polynomial time via, for instance, the Bellman-Ford algorithm [Bel58, CLR90].

```
ii ← 0
A: while ii < |V| do
  ii ← ii + 1
  // U(i, j) is the usage of the i-th resource class in control step j
  U(i, j) ← 0 for all i and j
  ℓ(v) ← 1 for all v ∈ V
  V' ← V
  while V' ≠ ∅ do
    // S is the set of unscheduled operations whose predecessors have been scheduled
    S ← {v|v ∈ V' and Pred(v) ∩ V' = ∅}
    choose v ∈ S in
      // try to find a control step l(v) where resources permit v to be scheduled
      for x in 1..ii do
        if U(r(v), l(v) mod ii) ≥ num(r(v)) then
          ℓ(v) ← ℓ(v) + 1
        endif
      endfor
      // if we have failed to schedule v in ii consecutive
      // steps, then it cannot be scheduled with ii as the
      // initiation interval. We increment ii and try again.
      if U(r(v), l(v) mod ii) ≥ num(r(v)) then
        continue A
      endif
      // Otherwise, we schedule v in cycle l(v)
      for each w ∈ Succ(v) do
        ℓ(w) ← max(ℓ(w), ℓ(v) + 1)
      endfor
      V' ← V' − {v}
      U(r(v), l(v) mod ii) ← U(r(v), l(v) mod ii) + 1
    endchoose
  endwhile
  // Check that all loop carried dependences are satisfied by the schedule
  for each loop carried dependence (v, v') ∈ E with distance d do
    if l(v) ≥ l(v') + ii * d then
      continue A // go back to outer loop, increment ii, and try again
    endif
  endfor
  return ii and l(v) for each v ∈ V // a feasible schedule has been found
endwhile
```

Figure 6.9 : Modulo Scheduling Algorithm

The *minimum initiation interval (MII)* is the max of the resource- and recurrence-constrained initiation intervals.

$$MII = \max(ResII, RecII) \tag{6.3}$$

The minimum initiation interval is, again, a lower bound on the value of a feasible initiation interval: certainly there are no legal schedules for smaller initiation intervals, but there is not necessarily a legal schedule for the MII, either. However, we can safely begin the search in Figure 6.9 with ii initialized to MII − 1 (so that ii will have value MII on the first iteration of the outer loop).

The following example illustrates modulo scheduling in the absence of recurrences.

Example 6.1. Consider the loop shown in Figure 6.10(a) and its dependence graph as shown in Figure 6.10(c). From the dependence graph one observes that there are no loop carried dependences in the given loop. Therefore, in this case *MII* = *ResII*.

The number of available resources (adders, etc.) per control step is given in Figure 6.10(b). From Equation 6.1, ResII for the given resource constraints is 2. The modulo scheduled loop is shown in Figure 6.10(d), with MII as the initiation interval. In Figure 6.10(d), control steps 1 and 2 correspond to the *prologue* of the transformed loop; the prologue is executed once when the modulo scheduled loop is started. Control steps 3 and 4 constitute the *kernel* of the transformed loop; the kernel is the repeating pattern that is executed over and

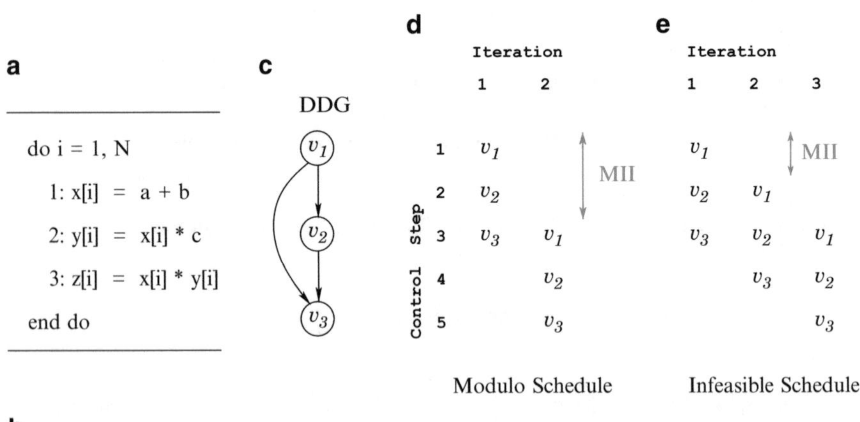

Figure 6.10 : Modulo scheduling in the absence of recurrences

over by the modulo scheduled loop, in this case executing operation v_3 from iteration i and operations v_1 and v_2 from iteration $i+1$. Control step 5 is the *epilogue* of the loop, which is executed just once to complete the final iteration before the loop terminates.

From the schedule one observes that 1 cycle is saved per iteration (except the first iteration) in the modulo schedule as compared to the sequential execution of the entire loop. Further overlap of iterations is not feasible as it would violate the resource constraints, e.g., in Figure 6.10(e), operation v_3 of the first iteration and operation v_2 of the second iteration cannot be scheduled in the same control step as there is only one multiplier available.

In Figure 6.9, the set S is the *ready set*, the set of operations where all dependence predecessors have been scheduled but the ready operations themselves have not yet been scheduled. Modulo scheduling involves selection of an operation amongst the operations in the ready set. Thus, like list scheduling, Algorithm 6.1 represents a family of algorithms parameterized by the method for selecting an operation from the ready set.

In the presence of *multi-function resources*, which can execute more than one type of operation, determining ResII becomes non-trivial. In such cases, a bestResII can be computed by performing an optimal bin-packing of the reservation tables for all the operations. The drawback of such an approach is its exponential complexity.

Next, we present an example of modulo scheduling in the presence of recurrences.

Example 6.2. Consider the loop shown in Figure 6.11(a) and its DDG as shown in Figure 6.11(b). From Figure 6.11(b) one observes that the dependence graph contains one cycle, $c = \langle v_2, v_5, v_6, v_7, v_2 \rangle$.

The length of the cycle ($len(c)$) is 4 and its distance ($dist(c)$) is 1. First, one computes the lower bound on the minimum initiation interval. Assuming the number of resources given in Figure 6.10(b), ResII for the DDG given in Figure 6.11(b) is 3 and RecII is 4. From Equation 6.3, one gets $MII = 4$. A corresponding schedule of the loop body is shown in Figure 6.11(c). From Figure 6.11(c), note that the delay of cycle c is 5 ($= del(c)$). Since $del(c) > len(c)$, the check that loop-

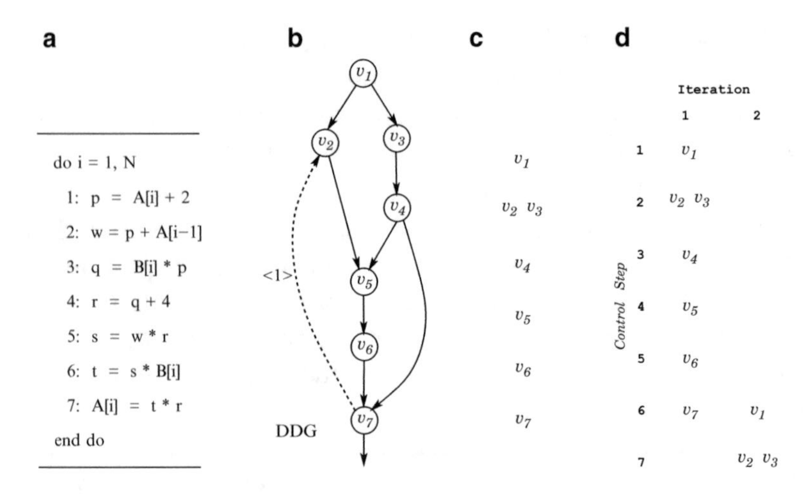

Figure 6.11 : Modulo scheduling in the presence of recurrences

carried dependences are preserved fails and so we determine that we have not been able to find a valid schedule with an initiation interval of 4. Therefore, the value of ii is incremented by one and the loop is rescheduled. The final modulo schedule is shown in Figure 6.11(d). As an exercise, the reader can verify that $ii = 5$ does not lead to any recurrence violation.

6.4.1 Remarks

Sufficiency of simple cycles

Algorithm 6.1 considers only simple cycles (defined in Section 6.3 on page 138) while computing RecII. However, it is not obvious that it is sufficient to consider simple cycles to guarantee a valid schedule. We illustrate the impact of both simple and non-simple cycles on RecII with the following example. Consider the dependence graph of Figure 6.12. The graph contains four simple cycles. Let $c_k = c_i \circ c_j$ denote concatenation of cycles c_i and c_j such that there is a traversal that includes all the nodes and c_k contains at least one operation twice (hence, c_k is not a simple cycle).

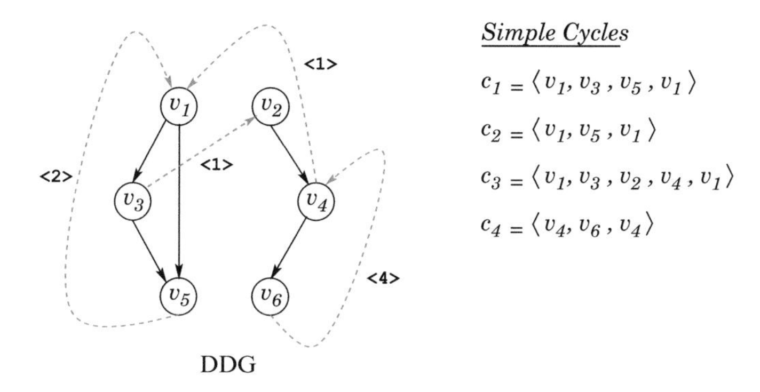

Figure 6.12 : Simple cycles in a dependence graph

From Figure 6.12 and Equation 6.2 (on page 142) , we observe that the simple cycle c_3 determines the value of RecII $(= 2)$ for the graph. Let us now consider the cycle $c_5 = c_3 \circ c_4$. The length of c_5 is 6; the distance of c_5 is 4. The recurrence constrained initiation interval for c_5 is computed as follows:

$$\left\lceil \frac{6}{6} \right\rceil = 1 < RecII$$

Thus, we observe that RecII for simple cycles is always greater than that of non-simple cycles.

Theorem 6.1. *Given a dependence graph $G(V, E)$, it is sufficient to consider only simple cycles for computing RecII.*

Proof. Consider two simple cycles c_i and c_j of $G(V, E)$. It is easy to see that RecII must satisfy:

$$RecII > \max\left(\frac{len(c_i)}{dist(c_i)}, \frac{len(c_j)}{dist(c_j)}\right)$$

Now, let us consider a cycle $c_k = c_i \circ c_j$. The length of c_k is $len(c_i) + len(c_j)$; the distance of c_k is $dist(c_i) + dist(c_j)$. The recurrence constrained initiation interval of c_k is given by

$$\frac{len(c_i) + len(c_j)}{dist(c_i) + dist(c_j)}$$

since

$$\max\left(\frac{len(c_i)}{dist(c_i)}, \frac{len(c_j)}{dist(c_j)}\right) > \frac{len(c_i) + len(c_j)}{dist(c_i) + dist(c_j)}.$$

Therefore, it is sufficient to consider only simple cycles for computing RecII. The proof also holds for cycles obtained by concatenating more than two cycles. Such cases can be reduced hierarchically to the case discussed above. For example, a cycle $c_4 = c_1 \circ c_2 \circ c_3$ can be hierarchically reduced to $c_4 = c'_1 \circ c_3$, where $c'_1 = c_1 \circ c_2$. □

Finding the Initiation Interval

As discussed previously, the MII is a lower bound on an initiation interval that will yield a feasible schedule. Rau et al. [RG81] used linear search to find an initiation interval starting at MII; we have also used this approach in Algorithm 6.1. The FPS compiler [Tou84] used binary search to find an initiation interval where the lower bound is MII and the upper bound is the length of the locally compacted loop body. A problem with binary search is that the existence of a schedule is not monotonic in the initiation interval: even if a certain ii has no feasible schedule, there may be an $ii' < ii$ where ii' has a feasible schedule. One could also employ an enumerative branch-and-bound search of all possible schedules or use a trial-and error approach guided by some heuristics. However, the former is computationally very expensive, while the latter does not always guarantee compact schedules. Huff [Huf93] modeled the problem as a minimal cost-to-time ratio problem [Law76]. Feautrier [Fea94], Govindarajan et al. [GAG94], Eichenberger et al. [ED95], Altman et al. [AGR95] model modulo scheduling as an integer linear programming [dD94] problem. However, the exponential complexity of integer linear programming algorithms render such techniques impractical for large loops. Similarly, techniques such as *simulated annealing*, *Boltzmann machine algorithm* and *genetic algorithms* [Ree93] have been proposed but can become very time expensive for large loops.

Because binary search cannot be guaranteed to find the minimum feasible initiation interval, and because of the expense of the more theoretically advanced techniques, modulo scheduling implementations generally use the straightforward linear search for a feasible initiation interval beginning at a lower bound such as MII.

6.4.2 Limitations

Modulo scheduling constrains each iteration to have an identical schedule and schedules successive iterations at a fixed initiation interval. However, in general, optimal schedules for many loops cannot be achieved by duplicating the schedule of the first iteration at fixed intervals. The constraint that each iteration has the same schedule is not necessary, but it is a fundamental component of modulo scheduling.

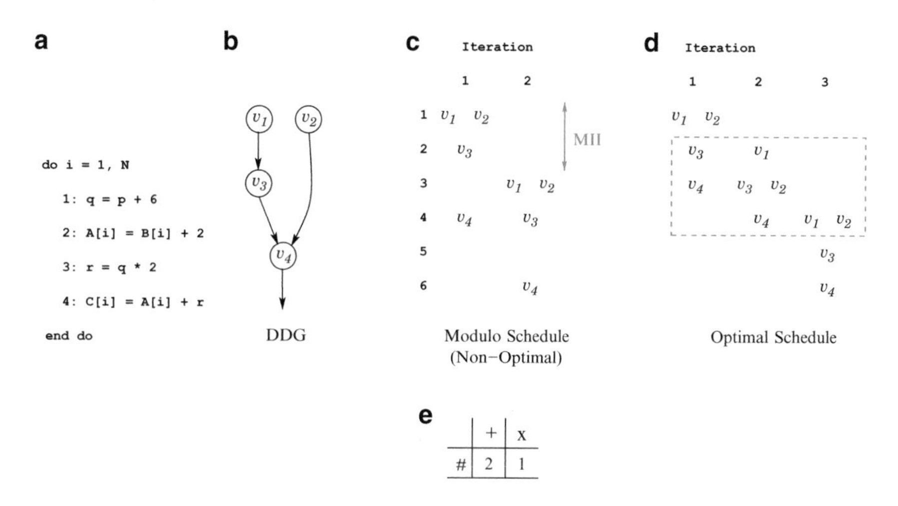

	+	x
#	2	1

Figure 6.13 : Limitations of modulo scheduling

For example, consider the loop and its data dependence graph shown in Figure 6.13(a) and Figure 6.13(b), respectively. Let v_i^k denote the i-th operation of the k-th iteration. The modulo schedule of the data dependence graph in Figure 6.13(b) is shown in Figure 6.13(c). Given resources shown in Figure 6.13(e), ResII for the dependence graph is 2. Since there are no loop carried dependences $MII = ResII$. Note that operation v_4^1 cannot be scheduled in control step 3 as it would lead to a resource conflict with operations v_1^2, v_2^2 scheduled in control step 3. Thus, a fixed initiation interval in conjunction with the modulo constraint can potentially create "gaps" in the schedule.

In order to extract higher levels of parallelism, one must somehow relax the modulo scheduling constraints and:

- Allow successive iterations to be scheduled at different initiation intervals.
- Allow iterations to have different schedules.

An optimal schedule is shown in Figure 6.13(d). Observe that the first two iterations have different schedules. Subsequently, iteration

3 is scheduled like iteration 1, iteration 4 is scheduled like iteration 2 and so on. Further, the second and the third iterations are initiated with different initiation intervals. The schedule of the second iteration in Figure 6.13(d), helps close the gap in the first iteration in Figure 6.13(c), as v_4^1 can now be scheduled in control step 3 without any resource conflicts. Thus, relaxing the constraints of modulo scheduling facilitates compaction which yields better performance.

As scheduling of iterations progresses, the execution of operations tend to form a repeating kernel, as shown in Figure 6.13(d) by the 'boxed' set of operations. The schedule in Figure 6.13(d) can be computed by modulo scheduling if the loop is unrolled once before modulo scheduling is performed. The difficulty for modulo scheduling is that the optimal kernel has different schedules for odd and even numbered iterations. But the optimal schedule for the loop unrolled once has the same schedule for every iteration of the unrolled loop body, since each iteration of the unrolled loop includes one odd and one even iteration of the original loop. The problem is that it is not clear how to predict how much unrolling is required for modulo scheduling to yield optimal schedules, as it may depend on resource constraints, data dependences, and the interaction between the two. While such kernels are problematic for modulo scheduling, we discuss a technique in subsection 7.3 that can directly detect such kernels. We further discuss the use of loop unrolling to improve modulo scheduling in subsection 6.7.2.

6.5 Modulo Scheduling with Conditionals

As originally described in [RG81], modulo scheduling assumes that the loop body does not contain conditional or unconditional jumps. In this section we discuss techniques for modulo scheduling loops with conditional branches in the loop body.

6.5.1 Hierarchical Reduction

Hierarchical reduction [Lam88] was proposed to make loops with conditionals amenable for modulo scheduling. The hierarchical reduction technique, derived from [Woo79], reduces a conditional to a *compound operation* whose scheduling constraints (both inter-iteration and loop carried dependences) represent the combination of the scheduling constraints of the two branches. That is, the set of operations that make up the conditional are grouped into a single operation. For each resource r, the requirement of the compound operation is the maximum of the use of r over the two branches. The latency of the compound operation is the latency of the longer branch. Finally, the dependences of the compound operation are all the dependences between constituent operations and other operations outside of the conditional.

The significance of hierarchical reduction is:

- Conditionals do not limit the overlapping of different iterations.
- It enables the movement of operations outside of a conditional around the conditional. Branches of different conditionals can be overlapped via hierarchical percolation scheduling (see Section 5.4.2).

The algorithm for hierarchical reduction can be summarized as follows: First, the `then` and `else` branches of a conditional statement are scheduled independently. Next, the entire conditional is reduced to a compound operation. In the presence of nested control flow, each conditional is scheduled hierarchically, starting with the innermost conditional.

The dependence graph thus obtained is free of conditionals and thus amenable to modulo scheduling using Algorithm 6.1. During code generation code is generated for both branches of a conditional. Any code scheduled in parallel with the conditional is duplicated in both branches.

Example 6.3. Consider the loop shown in Figure 6.14(a). The corresponding control flow graph is shown in Figure 6.14(b), where BB_i denotes the i-th basic block. From Figure 6.14(a), note that the operations v_2, v_3 and v_4 are dependent on operation v_1. Further, even though v_4 is in basic block BB_3, it can be scheduled in parallel with the conditional. As discussed above, the conditional, i.e., basic blocks BB_1 and BB_2, are replaced with the compound operation v'. The dependence graph of the transformed loop is shown in Figure 6.14(c).

The set of resources used by v' consists of an adder and a multiplier. Assuming the number of resources given in Figure 6.14(e),

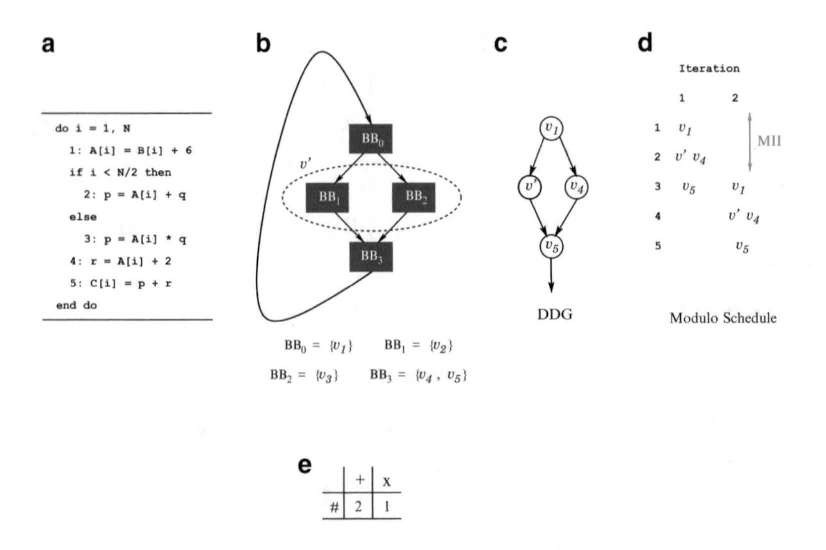

Figure 6.14 : Modulo scheduling with conditionals

modulo scheduling using Algorithm 6.1 results in the schedule shown in Figure 6.14(d).

6.5.2 Enhanced Modulo Scheduling

Predication (also known as *if-conversion*) is another technique for converting code with branches into data dependences, resulting in a single basic block [AKPW83, PS91]. A *predicated operation* consists of a predicate p and an operation o. The idea is that o is executed only if p is true. Typically the predicate is pre-computed, so that the execution of a predicated operation requires only checking one bit of state to determine whether the operation executes or not. For example, the conditional

if $x > 0$ **then** $x \leftarrow x + 1$ **else** $x \leftarrow 1$

could be written as the predicated code

true: $p \leftarrow x > 0$
p: $x \leftarrow x + 1$
$\neg p$: $x \leftarrow 1$

Predicated operations require special hardware support in the form of *predicate registers* to hold the condition for the operation. A predicate register is specified for each operation and predicate definition operations are used to set the predicate registers. Predication-based modulo scheduling techniques [RYYT89, DHB89] schedule operations along all execution paths together. Thus, the initiation interval for predicated execution is constrained by the resource usage of all the loop operations rather than those along the single execution path with maximum resource usage.

In the presence of multi-level branches, hierarchical predication [TLS90, MCH$^+$92, MCB$^+$93] or predicate matrices [MJ98] may be used to convert multi-level control dependences to data dependences. Early exit branches and multiple branches back to the loop header are handled in a similar fashion.

Enhanced Modulo Scheduling (EMS) [WBHS92] integrates hierarchical reduction with predicated execution to alleviate the limitations of each. Predication eliminates the need for pre-scheduling the conditional constructs, which is required in hierarchical reduction. Regenerating conditionals after modulo scheduling is complete eliminates the dependences on predicate registers assumed by predicated execution. Thus, EMS can be used on processors without support for predicated execution. We present an overview of the algorithm. First, the loop body (with conditionals) is converted into straight line predicated code using the RK algorithm [PS91]. The predicated loop body is then modulo scheduled using Algorithm 6.1 in conjunction with modulo variable expansion [Lam88] to rename registers with overlapping lifetimes. (Modulo variable expansion helps eliminate some data dependences, see subsection 6.7.1). Finally, the predicate definition operations are replaced with conditionals and predicated operations are placed in the appropriate branches.

Example 6.4. Consider the loop shown in Figure 6.15(a). The corresponding predicated code is shown in Figure 6.15(b). A predicate has an id and $type$, represented as a pair $\langle id, type \rangle$.[3] For example, operation v_4 in Figure 6.15(b) has a predicate with $id = p$ and $type = F$ (false—i.e., v_4 will execute if p is false). Operation v_3 sets the predicate if $x < N/2$ and clears the predicate otherwise. Operations that

[3] We follow the predicate notion proposed in [WBHS92].

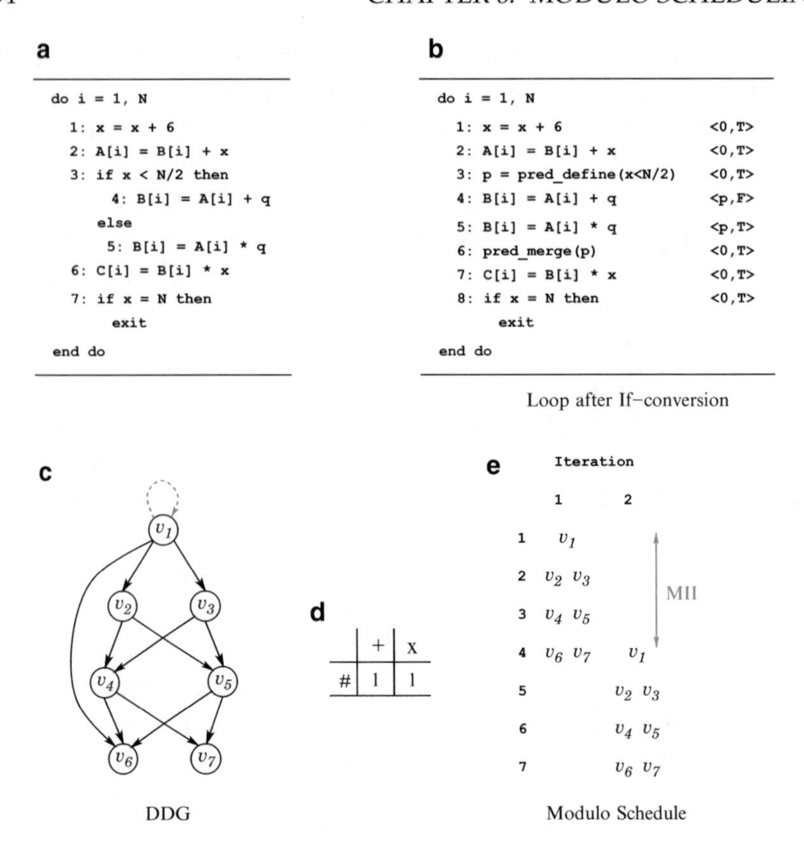

Figure 6.15 : Enhanced Modulo scheduling

are not dependent on a predicate such as v_1 are assigned the default predicate $\langle 0, type \rangle$. Operation v_7 is a loop bounds check and therefore does not define a new predicate.

The data dependence graph of the predicated loop is shown in Figure 6.15(c). Assuming the number of resources given in Figure 6.15(d), the MII for the predicated loop is 4. The predicated loop is then modulo scheduled, shown in Figure 6.15(e). Subsequently, conditionals are regenerated by replacing each predicate operation with a conditional; operations executed in parallel with a predicated operation are duplicated along both the branches of the corresponding conditional.

6.5.3 Modulo Scheduling with Multiple Initiation Intervals

A problem with hierarchical reduction is that if a conditional has branches of very different lengths then the initiation interval for the reduced loop body will be constrained by the worst path. Furthermore, the two paths of a conditional may have very different dependences (both intra-iteration and loop carried dependences). In general,

$$MII \geq \max_{\forall\, i} II_{path(i)} \tag{6.4}$$

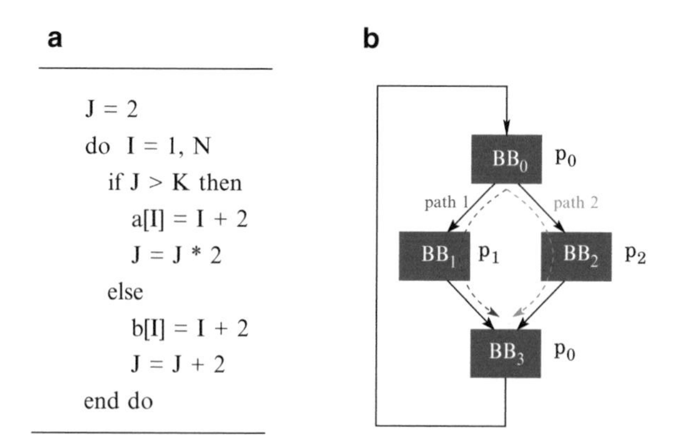

Figure 6.16 : A loop with control flow

where $II_{path(i)}$ is the initiation interval of path i. (Because hierarchical reduction aggregates resource and dependence constraints over all paths, note that the MII may be greater than the minimum initiation interval for any individual path.) From Equation 6.4 one observes that a path with a shorter initiation interval is penalized due to other paths. Similarly, when using predicated execution the MII is determined by the sum of the costs of all operations in the loop body. Thus, all paths are penalized due to the fixed MII. As an illustration, consider the example in Figure 6.16(a).

Assuming an availability of one adder and one multiplier, the MII of basic block BB_1 is 1 whereas the MII of basic block BB_2 is 2. A fixed MII using hierarchical modulo scheduling assigns an MII of 2 to both paths, which slows down the execution of path 1.

In [WP95], modulo scheduling with multiple initiation intervals was proposed for architectures with support for predicated execution. In Figure 6.16(b) the predicate for a basic block is shown next to the block. Predicate p_0 is the predicate `true` (i.e., operations predicated with p_0 always execute). After if-conversion, path 1 corresponds to the operations predicated on p_0 and p_1, and path 2 to the operations predicated on p_0 and p_2. The initiation intervals for the two paths is shown in Figure 6.17.

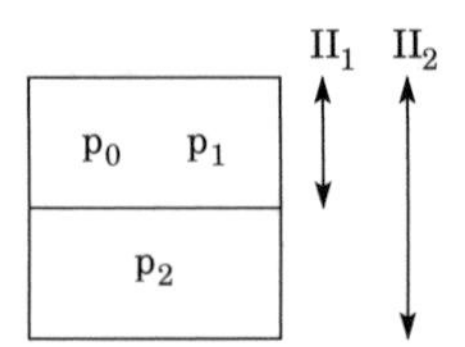

Figure 6.17 : Multiple initiation intervals

From Figure 6.17 one notes that during execution of path 1, only operations from path 1 are fetched and executed. However, during the execution of path 2, operations from both the paths are fetched and operations predicated on p_1 are squashed. Path 2 corresponds to a longer initiation interval II_2 due to the larger number of operations. Notice that path 1 is never penalized. The execution paths of a loop body often have different execution frequencies. In order to minimize penalty during the scheduling process, the algorithm assigns higher priorities to the more frequently executed paths, i.e., smaller initiation intervals are assigned to the more frequently executed paths. A detailed description of the algorithm can be found in [WP95].

6.6 Iterative Modulo Scheduling

In the modulo scheduling algorithms presented so far, operations are scheduled in some order and if at any point an operation cannot be scheduled the entire partial schedule is abandoned and the algorithm retries with a larger initiation interval. In this section we discuss *iterative modulo scheduling* [Rau95], a modulo scheduling algorithm that does not discard the schedule when it is discovered to be infeasible. The algorithm is iterative in the sense that it schedules and reschedules operations to find a schedule that simultaneously satisfies all the scheduling constraints. When the scheduler finds that there is no slot available for scheduling the current operation, it displaces one or more previously scheduled, conflicting operations. These are in turn re-scheduled as the search continues. If the search fails to yield a valid schedule even after a large number of steps, then the initiation interval is increased and the entire process is repeated.

6.6.1 The Algorithm

Each time an operation is scheduled, the iterative modulo scheduling algorithm *tentatively* schedules the same operation of subsequent iterations at intervals of *II*. The algorithm allows the unscheduling of other tentatively scheduled operations from subsequent iterations due to resource or dependence conflicts. When possible, however, operations are scheduled so as not to unschedule any operation that was tentatively scheduled previously. Note that since it is only a tentative schedule, it does not block the forward progress of the scheduler in the event that the current operation cannot be scheduled because it conflicts with one or more operations from a subsequent iteration.

Algorithm 6.2. Iterative Modulo Scheduling (IMS)
The input to this algorithm is a dependence graph $G(V, E)$ of a loop $\mathcal{L}$ and a user-defined parameter *Budget* giving the maximum number of operations scheduled before giving up and trying a larger initiation interval. The dependence graph has two distinguished nodes - *start* and *stop* - which represent the entry and exit points of the graph. As before, the MII is determined using Equation 6.2 (on page 142). For simplicity, unscheduling is omitted from the pseudo-code, but is discussed below.

// Initialize *II*
II $\leftarrow$ *MII*
isScheduled $\leftarrow$ *false*
while $\neg$ *isScheduled* **do**
  Mark all operations unscheduled
  Insert all operations in the list of unscheduled operations
  for each $v \in V$ **do**
    Compute height-based priorities
  endfor
  Schedule *start* at time 0
  Budget $\leftarrow$ *Budget* -1
  while *all operations not scheduled* and *Budget* $\neq 0$ **do**
    // Pick the highest priority operation v_h

```
      v_h = HighestPriorityOperation()
      // Determine earliest start time, ASAP(v_h)
      ASAP(v_h) = CalculateEarlyStartTime(v_h)
      ALAP(v_h) = ASAP(v_h) + II - 1
      // Find time slot to schedule v_h
      t = FindTimeSlot (v_h, ASAP(v_h), ALAP(v_h))
      Schedule v_h at time t  (*)
      Budget <- Budget -1
   endwhile
   if all operations scheduled then
      isScheduled <- true
   else
      Set Budget to its initial value
      II <- II +1
   endif
endwhile
```

Like other modulo scheduling algorithms, IMS maintains the invariant that at each step all resource and dependence constraints are satisfied for the operations in the partially constructed schedule. Thus, if the algorithm reaches a point where all the operations from one iteration have been scheduled (without having to displace any tentatively scheduled operations from the subsequent iterations) then, by construction, the complete data dependence graph has been modulo scheduled. The repeating portion of the resulting schedule forms the body of the new loop; this loop is preceded by a prologue and followed by an epilogue which set up and complete correct execution of the new loop. This is a special case of software pipelining using kernel recoginition, which we describe in detail in Chapter 7.

Unlike traditional acyclic list schedulers, IMS may also *unschedule* an operation after it has been scheduled. Specifically, assume an operation o cannot be scheduled on line $(*)$ in the algorithm given above due to resource or dependence constraints with already-scheduled operations. In this situation IMS schedules o anyway at the selected time. If scheduling o violates a data dependence with operation o', then o' is unscheduled (i.e., removed from the schedule). If scheduling o violates a resource constraint with some set of operations $o_1, o_2, \ldots$ that all use that same resource at the same time, then one of the o_i is unscheduled to make room for o. Any operation that is unscheduled is put back on the list of unscheduled operations with its original priority.

Determining Scheduling Priority

Standard list scheduling techniques [ACD74] employ a *height-based* priority function [Hu61, RCG72], which has two important properties:

- If there is a dependence between two operations, then the priority function gives the predecessor operation a higher priority than the successor operation.
- Among operations that are independent, higher priority goes to operations with smaller *slack*, which is the difference between the ALAP and ASAP labels of an operation (see Chapter 3). Intuitively, an operation o with small slack has a narrow range of time steps in which it can be placed without lengthening the overall schedule and so o should be given high scheduling priority once all of o's dependence predecessors are scheduled.

The conventional height-based priority works only on acyclic dependence graphs and so must be extended in the context of IMS to account for loop carried dependences.

Definition 6.3. *A strongly connected component (SCC) of a directed graph* $G(V, E)$ *is a maximal set of vertices* $U \subseteq V$ *such that every pair of vertices in* U *are reachable from each other.*

IMS partitions the data dependence graph into a set of SCCs. Each SCC in the graph is replaced by a *super node*. Hence, the resulting graph obtained is acyclic. The height of all nodes (both normal nodes and super nodes) is computed by a depth-first walk of this acyclic graph. After this computation, the heights of the vertices in each SCC are computed iteratively by a post-order visit beginning at an entry point to the SCC. Consider a successor v of operation u with a distance of $D(u, v)$ (recall that the distance between two operations is the number of iterations separating them). Let the latency between the operations u and v be denoted by $\mathcal{D}(u, v)$. Assume that the operation u that is in the same iteration as v (the current iteration) has a height-based priority h. Since, u's successor v is actually $D(u, v)$ iterations later, its height-based priority relative to current iteration is effectively given by:

$$\mathcal{h}(u) = \begin{cases} 0 & \text{if } u = \textit{stop}, \\ \max\limits_{v \in \text{Succ}(u)} (\mathcal{h}(v) + \mathcal{D}_{\text{eff}}(u, v)) & \text{otherwise.} \end{cases} \quad (6.5)$$

where $\mathcal{h}(u)$ is the height of an operation u, *Succ(u)* is the set of successors of u and $\mathcal{D}_{eff}(u, v)$ (effective latency between operation u and v) is given by:

$$\mathcal{D}_{\text{eff}}(u, v) = \mathcal{D}(u, v) - II \times D(u, v)$$

Determining Earliest Start Time

The function *CalculateEarlyStartTime* determines the earliest time an operation can be scheduled as constrained by its predecessors. The calculation of *ASAP(v)* is affected by the fact that operations can be unscheduled—it may be that some of the predecessors of v are not scheduled when v is scheduled. We modify $ASAP(v_h)$ so that it can be calculated by considering only the scheduled immediate predecessors of the operation. The earliest time an operation v can be scheduled is given by :

$$\text{ASAP}(v) = \max_{u \in \text{Pred}(v)} (\text{ASAP}'(u) + \mathcal{D}(u, v)) \quad (6.6)$$

where $\text{ASAP}'(u)$ is given by :

$$\text{ASAP}'(u) = \begin{cases} 0, & \text{if } u \text{ is unscheduled} \\ \max(0, \text{SchedTime}(u) + \mathcal{D}_{\text{eff}}(u, v)) & \text{otherwise} \end{cases}$$

where, *SchedTime(u)* is the time at which operation u is scheduled.

Determining Candidate Time Slots

In iterative modulo scheduling, it is impossible to guarantee that all the predecessors of an operation o have been scheduled and have remained scheduled. That being the case, dependences with the predecessor operations are honored by not scheduling o before ASAP(o). To honor dependences with successor operations, there is also a latest time by which o must be scheduled. However, the two conditions may be incompatible—there may be no time at which o can be scheduled without violating predecessor, successor, or resource constraints. In this situation, o is scheduled in a time step that satisfies predecessor constraints but may violate successor or resource constraints. Any successor constraints that are violated cause that successor operation to be unscheduled. For resource conflicts, there may be several operations that use the same resource; the lowest priority operation is selected.

6.7 Optimizations

In this section we discuss auxiliary optimizations that can play a large role in extracting parallelism from loops. Note that these optimizations can benefit trace scheduling as well, but have the greatest impact on loop parallelization.

6.7.1 Modulo Variable Expansion

(a) Original Schedule	(b) Schedule after Modulo Variable Expansion

<pre>   do i=1, N
1:    x = 3*i
2:    a[i] = i
3:    b[i] = x+1
    end do</pre>
	<pre>x1 = 3*i
	a[i] = i x2 = 3*i
	b[i] = x1+1 a[i] = i x1 = 3*i
	                b[i] = x2+1 a[i] = i
	                                b[i] = x1+1</pre>

Figure 6.18 : Modulo Variable Expansion

Modulo variable expansion [Lam88] applies a static version of a traditional register renaming technique to remove some inter-iteration data dependencies. Consider the example in Figure 6.18(a) where a value is written into a register and used two cycles later. Assuming we have one adder and one multiplier available, the *ResII* is 1. However, due to the loop carried anti-dependence (recall Definition 3.1) on x a new iteration cannot be started before operation 3 of the preceding iteration (refer to Figure 6.18(a)) has completed. Therefore, the throughput is limited to one iteration every three cycles. The code can be sped up using different registers for the variable x in alternating iterations, as shown in Figure 6.18(b). The transformation identifies any variable such as x with a loop carried anti-dependence and "expands" that variable into multiple variables so that each iteration can refer to a different location. By construction, the uses of the variable(s) in overlapping iterations are independent, and the loop can be pipelined with a shorter *II*. Modulo variable expansion minimizes renaming, and thus the number of additioinal memory locations required, by reusing the locations in non-overlapping iterations. Also, modulo variable expansion in association with register pipelining reduces memory traffic, and hence improves program execution time significantly.

6.7.2 Using Loop Unrolling to Enhance Modulo Scheduling

Modulo scheduling requires an integral initiation interval. The rounding of the initiation interval to an integer (recall Equation 6.4 on page 155) limits overlapping of operations from different iterations than is possible otherwise, thereby leading to performance degradation. Furthermore, the initiation interval is restricted to a minimum value of 1, limiting the performance of a modulo scheduled loop to one iteration per cycle. Loop unrolling (recall Section 6.2 on page 135), before applying MS, helps to mitigate the effect of rounding of the initiation interval by creating larger loop bodies which require more resources, thus increasing ResII [LH95]. The larger the ResII, the smaller the impact of rounding up the initiation interval to the next integer. Similarly, RecII may not be an integer if the length of a cycle is not an integral multiple of the distance of the cycle. In such cases, loop unrolling is performed to reduce the distance of the recurrence.

Unrolling facilitates (upward) percolation of operations from successive iterations as the operations are free of the iteration boundaries. Further, unrolling increases the rate at which iterations are executed as illustrated in the following example. Thus, in effect, a non-integer initiation interval is achieved as more than one iteration of the original loop is initiated during the same initiation interval.

Example 6.5. Consider the loop and its dependence graph shown in Figure 6.19(a) and Figure 6.19(b) respectively. Assuming the number of resources given in Figure 6.19(f), the corresponding modulo

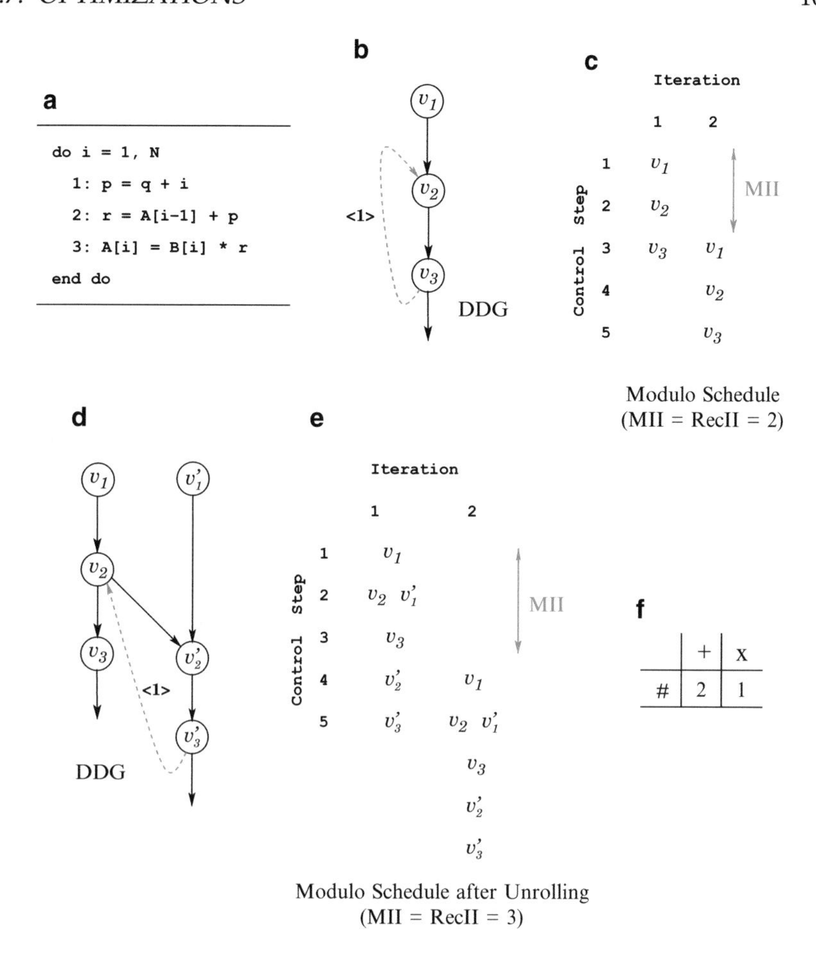

Figure 6.19 : Modulo scheduling with unrolling

schedule is shown in Figure 6.19(c) where successive iterations are initiated every two control steps.

In order to achieve a smaller initiation interval, we unroll the loop once (i.e., add a second copy of the loop body to the loop). The dependence graph of the unrolled loop body is shown in Figure 6.19(d); operations, denoted by v'_i, correspond to the new operations in the loop body, introduced as a result of loop unrolling. The modulo schedule of the unrolled loop is shown in Figure 6.19(e). From Figure 6.19(e) we observe that two iterations of the original loop are initiated every three control steps. Therefore the effective initiation interval is 1.5. Thus, unrolling the loop prior to modulo scheduling reduces the initiation interval.

Additionally, unrolling may enable other optimizations which can reduce the resource requirements and dependence height [MCGH92]. Though loop unrolling exposes higher levels of instruction level parallelism, it must be used carefully. Loop unrolling may result in an increase in instruction cache misses and register pressure,[4] which in the worst case can more than cancel the expected benefit of the transformation. Note that the gain in performance decreases progressively with increasing code size. Heydemann et al. [HBK+03] proposed the *UFC (Unrolling Factor computation under Constraints)* method to compute unrolling factors of a set of loops while taking into account code size.

FURTHER READING

A comparative study of the various modulo scheduling techniques is presented in [CLG02, KN11]. Modulo scheduling of control-intensive loops in non-numerical programs is discussed in [LH96]. A cache-aware modulo scheduling technique is discussed in [SG97]. A clustered modulo scheduling technique for VLIW architectures with distributed cache is discussed in [SG01]. Ding et al. proposed a new modulo scheduling technique to boost cache reuse [DCS97]. Code generation schemas for modulo scheduled loops are discussed in [RST92]. In [LF02], Llosa and Freudenberger discuss a reduced code size modulo scheduling technique; Merten and Hwu proposed a new architectural mechanism, called *Modulo schedule buffers*, to contain code expansion during modulo scheduling [MH01].

Clustered VLIW architectures are typically characterized with clusters of a few functional units and small private register files. In [NE98], Nystrom and Eichenberger proposed a cluster assignment for modulo scheduling; other approaches for cluster assignment during modulo scheduling include [ACSG01, ACS+02]. Instruction scheduling in general for clustered VLIW architectures is discussed in [SG00b]. Various modulo scheduling approaches have been proposed for such architectures [FLT99, SG00c]. The effectiveness of loop unrolling for modulo scheduling in the context of clustered architectures is discussed in [SG00a].

A large number of approaches have been proposed over the last three decades for loop pipelining. In [GS94], Gasperoni and Schwiegelshohn propose an algorithm

[4] The longer unrolled loop body may result in different register allocation that increases register pressure.

for scheduling loops on parallel processors. In [WE93] Wang and Eisenbeis proposed a technique called *Decomposed Software Pipelining* wherein software pipelining is considered as an instruction level transformation from a vector of one-dimension to a matrix of two-dimensions. In particular, the software pipelining problem is decomposed into two subproblems—one is to determine the row-numbers of operations in the matrix and another is to determine the column-numbers. In [RGSL96], Ruttenberg et al. compared optimal and heuristic methods for modulo scheduling. Allan et al. surveyed various techniques for software pipelining in [AJLA95].

Several techniques [JPSW82, GFO92, Ree93] have also been proposed to find the global minimum of the search space of feasible schedules. However, the high complexity of such approaches makes them currently very expensive in practice.

7

SOFTWARE PIPELINING BY KERNEL RECOGNITION

Kernel recognition techniques avoid the search for an appropriate initiation interval by dealing directly with a representation of the unrolled loop and its compaction. Intuitively, kernel recognition tries to achieve the effect of fully unrolling and then compacting the loop. In this chapter we present a number of kernel recognition methods that overcome the limitations of modulo scheduling in different ways, including methods that generate provably optimal schedules (in the absence of resource constraints) and handle the scheduling of tests directly, without the need for an intermediate abstraction such as hierarchical reduction.

7.1 Introduction

Kernel recognition techniques avoid the search for an appropriate initiation interval by dealing directly with a representation of the unrolled loop and its compaction. Intuitively, kernel recognition tries to achieve the effect of fully unrolling and then compacting the loop. Of course, it is not a priori obvious that this is an achievable goal, in the sense that there is no reason to believe that the parallelism in every loop (or indeed any loop) can be fully captured by any closed (fixed size) representation independent of full unrolling and compaction of the loop. Indeed, a simple counterexample is given in Figure 7.1. In this loop the iterations are independent of each other, and therefore

A. Aiken et al., *Instruction Level Parallelism*,
DOI 10.1007/978-1-4899-7797-7_7

```
do i = 1, N
    a[i] = 0.0
end do
```

Figure 7.1 : Loop where amount of parallelism is proportional to the number of iterations

no fixed size unrolling can capture all of the parallelism: each extra iteration added will bring in additional parallelism. Thus, in this case a closed form expression of the parallelism at the ILP level is not feasible.[1]

It turns out, however, that trivially parallelizable loops such Figure 7.1 are the only kind of loop that cannot be handled naturally by kernel recognition techniques. Indeed, in some cases it is possible to show that provably all of the ILP parallelism is extracted by a kernel recognition algorithm.

The goal of any software pipelining algorithm is to identify a new loop body that can be substituted for the original loop body. When compacting an unrolled loop, this new loop body manifests itself as a repeating pattern of parallel instructions. Since, as just discussed, such a pattern does not always emerge naturally, software pipelining techniques force the occurrence of such a pattern by imposing various constraints on the schedule. For example, modulo scheduling imposes two constraints (an identical schedule for each iteration and a fixed initiation interval) which together are sufficient to guarantee a repeating pattern if scheduling succeeds. While all techniques for exploiting ILP in single loops (including loop unrolling) share the above overall approach, they differ greatly in the type of constraints they impose to ensure that a repeating pattern will emerge. For example, some techniques disallow conditional branches in the loop body and/or use resource constraints to ensure convergence (termination) of the software pipelining process. These restrictions in turn influence the generality of the technique, its effectiveness in extracting parallelism, and its efficiency. For example, in modulo scheduling, a fixed

[1]This loop can, of course, be expressed as a vector loop. But that does not capture the explicit nature of the ILP parallelism and is not a closed form pattern in the sense used here, since the vector instruction itself would be parameterized by N or be broken up in smaller chunks which do not (directly) capture the intrinsic parallelism of the whole loop.

initiation interval limits the exploitation of parallelism, as discussed in Section 6.4.1. The techniques in this chapter often expose and exploit more parallelism, but they also tend to require more time and space to do so.

7.1.1 Basic Idea

The basic idea behind kernel recognition is to unroll the loop and compact operations from different iterations "as much as possible"[2] while looking for a repeating block of instructions (a *"pattern/kernel"* to emerge). This repeating pattern constitutes the loop body of the new loop with a corresponding prologue and an epilogue which, respectively, initiate and complete the kernel. An example is shown in Figure 7.2. In this example, the first several iterations of the loop are unrolled and operations are scheduled greedily as early as possible. The dashed box represents the kernel of the software pipelined loop: If we continue to add more iterations and schedule them using the same greedy policy, the pattern within the dashed box will simply repeat itself. The prologue is all the code before the first occurrence

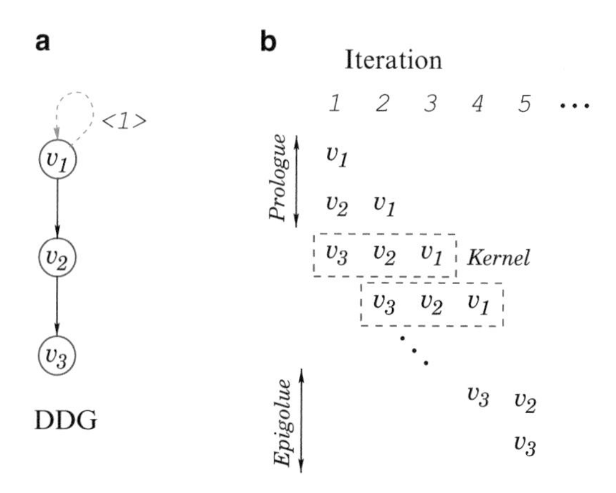

Figure 7.2 : Kernel identification during software pipelining

[2]In the presence of vectorizable operations in the loop body, as in Figure 7.1, the compaction of operations must be constrained in some way to force the emergence of a repeating pattern. A simple solution is to not allow not more than a fixed number of iterations to fully overlap, i.e., there must be a delay of at least one control step, even if there are no data dependences that require such a delay.

of the pattern and the epilogue is all the code after the last occurrence of the pattern.

Notice that in this case the kernel arises naturally, in the sense that further unrolling and scheduling result in the repetition of the same pattern: the parallel execution of operations v_1 from iteration $i + 2$, v_2 from iteration $i + 1$, and v_3 from iteration i. Intuitively, this kernel is what we want to capture as the loop body. If the kernel contains parts of n iterations executing concurrently, the prologue must start $n-1$ iterations ahead preparing for the steady state kernel to take over and the epilogue must finish $n - 1$ iterations from the point where the steady state kernel leaves off. Note that the example above is simplified to give an intuitive understanding of kernels, and we are ignoring details such as the loop increment, exit test, and operation latencies. Handling such features does not impose any conceptual difficulty and will be dealt with in the subsequent sections.

The rest of the chapter is organized as follows. In Sections 7.2–7.3, we present some well-known techniques for software pipelining using kernel recognition. In Section 7.4 we extend these techniques to handle conditional statements. We conclude by discussing extensions to nested loops (Section 7.5) and procedure calls (Section 7.6).

7.2 The URPR Algorithm

Su et al. proposed the *URPR* (Unrolling, Pipelining and Rerolling) algorithm [SDX86], an extension of the microcode loop compaction algorithm *URCR* [SDX84], for software pipelining of innermost loops. It assumes that the iteration count is known statically and that the loop body has no conditionals or procedure calls. The URPR algorithm is outlined below.

Algorithm 7.1. UnRolling, Pipelining and Rerolling
The input to the algorithm is an acyclic flow graph $G(V, E)$ that represents the body of a loop $\mathcal{L}$. The output is a software pipelined version of the loop $\mathcal{L}$.

// Compact the loop body

Schedule $G(V, E)$ using Algorithm 3.1

Let L denote the length of the compacted loop body

Let D be the maximum distance of all the simple cycles in $G(V, E)$

// Compute the unroll factor

$K = \lceil \frac{L}{D} \rceil$

Unroll the loop body K times

Pipeline the unrolled loop body $G'(V', E')$ (see below)

Reroll the pipelined loop body

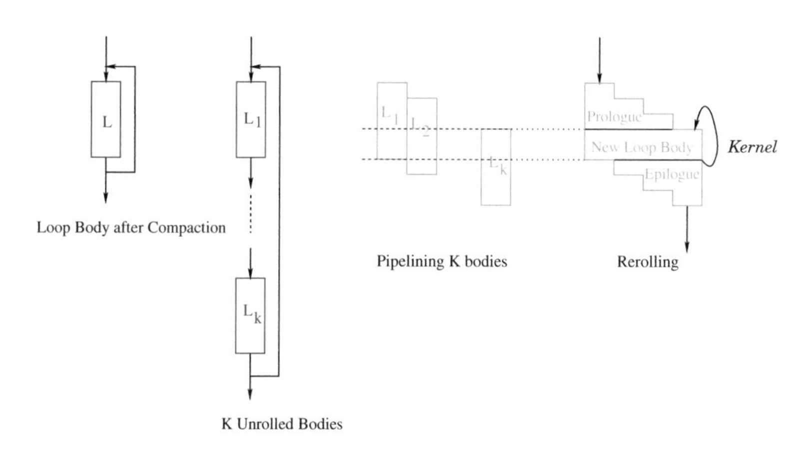

Figure 7.3 : The URPR Algorithm

The different stages of the URPR algorithm are illustrated in Figure 7.3. First, URPR compacts the loop body and then it unrolls the loop, which introduces new operations in the loop body (which in turn requires the dependence graph for the unrolled loop body to be reconstructed). The iterations of the unrolled loop (with identical schedules) are pipelined subject to the loop carried dependences and resource constraints. Loop rerolling finds the kernel of the software pipelined loop that includes all the operations in the original loop body and has the shortest length. Finally, the back edge around the original loop body is moved to go around the kernel. Since the order of compacted operations is not altered during the pipelining phase, i.e., the successive iterations of the original loop body have identical schedules, a kernel will always emerge.

The selection of the optimal unrolling factor (K) is critical for pipelining and rerolling of the unrolled loop bodies. Too small a K

restricts the formation of a new loop body, possibly limiting parallelism whereas too big a K does not improve the final rerolled loop. Unrolling for optimal pipelining is discussed in [JA90]. In the following example we illustrate Algorithm 7.1.

Example 7.1. Consider the dependence graph shown in Figure 7.4(a). The corresponding URPR schedule is shown in Figure 7.4(b). Note that the distance (in number of control steps) between operation v_3

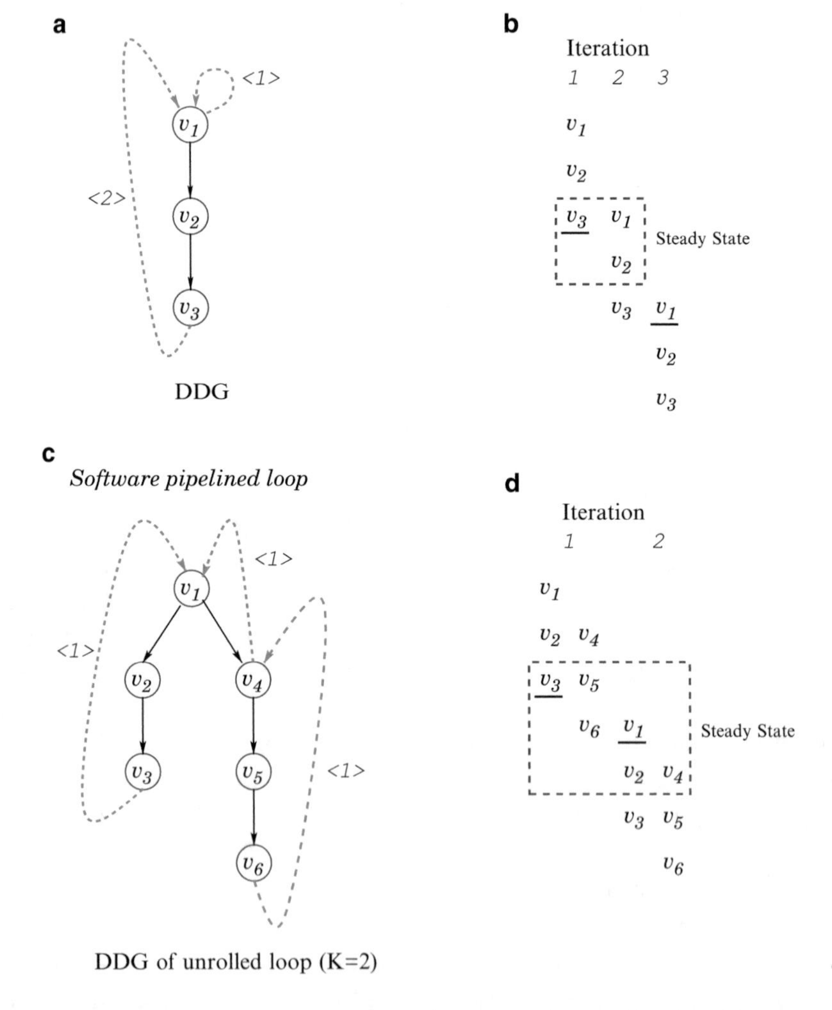

Figure 7.4 : Unrolling and pipelining in the URPR algorithm

of the first iteration and operation v_1 of the third iteration is 2. From Figure 7.4(a), we observe that the length of the loop L is 3 and the maximum dependence distance D is 2. Therefore, the unroll factor $\lceil L/D \rceil$ is 2. Therefore, the loop is unrolled two times. The dependence graph of the unrolled loop is shown in Figure 7.4(c). Note that in the unrolled loop body, operation v_4 (resp. v_5 and v_6) is the same as operation v_1 (resp. v_2 and v_3) of the second iteration.

The unrolled iterations are then pipelined subject to the loop carried dependences and resource constraints. The final schedule loop is shown in Figure 7.4(d). Observe that the number of control steps between the underlined operations in Figure 7.4(b) is 2, whereas the number of control steps between the same underlined operations in Figure 7.4(d) has improved to 1. Note that the schedule shown in Figure 7.4(b) initiates a new iteration in 2 steps, whereas the URPR schedule in Figure 7.4(d) initiates two iterations every 3 steps. Thus, unrolling the loop prior to scheduling exposes higher levels of parallelism. The schedule in Figure 7.4(d) cannot be further compacted due to the dependence between operation v_3 of the first iteration and operation v_1 of the second iteration.

Note that although pre-compaction, i.e., compaction before unrolling, makes kernel identification simpler, it can potentially delay the execution of dependence-free operations of the successive iterations and thus, can restrict the amount of parallelism. A simple example is in Figure 6.19(c) where the execution of operation v_1 of the second iteration is delayed to the third control step, whereas post-compaction enables the execution of the same operation in second control step, as shown in Figure 6.19(e). An integrated approach where compaction and pipelining are done "on the fly" is discussed in Section 7.4.1.

Su et al. extend URPR to GURPR (Global URPR) [SDX87, SW91] to handle loops with abnormal entries, conditional exits, more than one path, and with nested loops and procedure calls. GURPR first performs global scheduling on the loop body and then employs URPR, assuming if-conversion [AKPW83] and hierarchical reduction of inner loops, to derive a repeating pattern. The shortcomings of URPR discussed above are inherited by GURPR.

7.3 OPT: Optimal Loop Pipelining of Innermost Loops

As discussed at the beginning of Chapter 6, a simple approach to extracting ILP from loops is to unroll the loop before compaction. If the loop is unrolled (say) k times then parallelism can be exploited in this unrolled loop body by compacting the k iterations, but the new loop still imposes sequentiality between every group of k iterations. Although additional unrolling and compaction can be done to expose higher levels of parallelism, this becomes expensive very rapidly. OPT [AN88] overcomes this problem by achieving the effect of unbounded unrolling and compaction of a loop.

The schedule obtained using OPT is *time optimal*, i.e., no transformation of the loop based on the given data dependences can yield a shorter running time for the loop. The optimality result depends on being able to schedule any number of operations in a single cycle. In the presence of resource constraints, a simple extension of the algorithm guarantees a schedule within a factor of two of optimal and normally performs much better than that in practice.

There is an interesting contrast with the design of modulo scheduling. Modulo scheduling deals directly with data dependences and resource constraints, but must search for a feasible initiation interval. OPT deals directly with the data dependences and the initiation interval, but handles resources in a separate step.

The basic technique examines a partially unrolled loop, say the first i iterations, and schedules the operations of those iterations greedily as soon as possible, as in the example in Figure 7.2. However, in more complex codes the greedy approach may introduce *gaps*—control steps with no operations—between the operations of an iteration due to some operations having loop carried dependences with greater distance than others. With greedy scheduling, these gaps may grow with each successive iteration and in turn prevent formation of the repeating pattern needed for a kernel. In such cases, the algorithm shrinks the gaps, i.e., the operations are rescheduled such that there is no effect on the critical path length but the operations issue at the same rate as the critical path, which is sufficient to guarantee the emergence of a kernel. A kernel can be inferred from a greedy schedule where gap sizes, if any, are bounded by a constant, in at most $O(n^2)$ iterations, provided all cycles in the dependence graph have distance one [AN88].

We next define some terms and present the algorithm for optimal parallelization of innermost loops. Let v_i^k be the i-th operation of the k-th iteration.

Definition 7.1. *The* **slope** *of cycle c in a dependence graph is the ratio $len(c)/dist(c)$ [CCK88].*

The slope of a cycle establishes a bound on the rate that operations in the cycle can be executed. Let *slope*(v) be the maximum slope of any cycle on which v depends. If v is not dependent on any cycle, then *slope*$(v) = 0$. When an iteration is scheduled, it is spread across some interval of control steps $t_1, \ldots, t_k$. Operations from an iteration tend to cluster into groups of mutually dependent operations (referred to as a *region*) with gaps between the regions. For example, in Figure 7.5(b) there are two maximal regions.

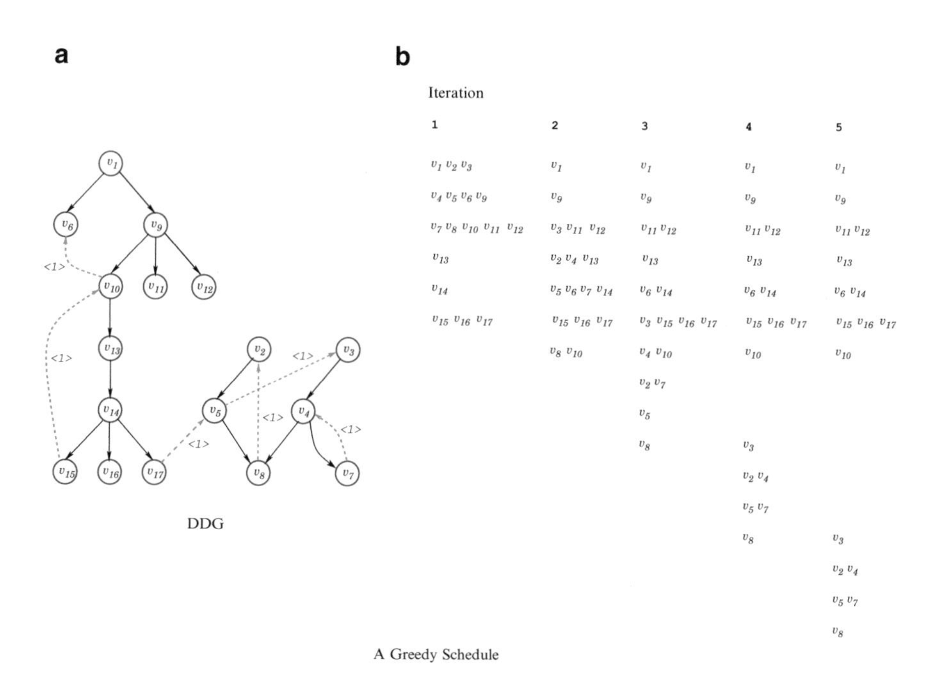

Figure 7.5 : A Greedy Schedule

Definition 7.2. *Let $A_1, \ldots, A_j$ be the maximal regions of an iteration, where A_i spans time steps $t_i, \ldots t_{i'}$ and $t_{i'} < t_{i+1}$ for all i. Then* **gap**$(A_i, A_{i+1}) = t_{i+1} - t_{i'}$.

In Figure 7.5(b), iterations four and five have the same maximal regions; the only difference between the iterations is the size of the gap. Two iterations i and $i + c$ are said to be *alike* if they have the same maximal regions in the same order and the gaps in iteration $i + c$ are as large or larger than in iteration i.

The last region A_j of an iteration can never be scheduled earlier because every operation is already scheduled as early as possible. Thus, the only way the gaps in an iteration can be shrunk is by rescheduling regions A_k where $k < j$ later—i.e., closer to A_j. Observe that it is always legal to shrink the gaps in the last (but only in the last) currently scheduled iteration. In fact, the gaps can be closed completely by rescheduling region A_{j-1} to end in the control step immediately before the start of region A_j, rescheduling A_{j-2} to end in the control step immediately before the start of region A_{j-1}, and so on. However, while simply closing the gaps in the last iteration scheduled so far preserves dependences, it does not necessarily preserve optimality because closing the gaps may cause operations on the critical path in future iterations to be delayed.

In [AN88] a sufficient condition for closing gaps and preserving time optimality is given. The gaps in iteration $i + c$ can be shrunk to the size of the gaps in iteration i if there is no *unbroken* dependence path from an operation in A_k of iteration i to an operation in A_j of iteration $i + c$, where iterations i and $i + c$ are alike with j maximal regions and $k < j$. A dependence path is unbroken if each operation in the chain is scheduled in the control step immediately after its predecessor completes. If every dependence chain from region A_k in iteration i to region A_j in iteration $i + c$ is broken, then shrinking gaps so that iteration $i + c$ is identical to iteration i cannot delay any operation in region A_j of iteration $i + 2c$. Since A_j contains all the operations on the critical path in iteration $i + 2c$, there is still an optimal schedule even with the gaps in iteration $i + c$ shrunk to the gap sizes in iteration i.

For example, in Figure 7.5(b) iterations four and five are alike and there is no unbroken dependence path from the first region of iteration four to second region of iteration five. Thus, it is safe to shrink the gap in iteration five to the size of the gap in iteration four. Though in Figure 7.5(b) the gaps can be completely closed, it is not always safe to completely close the gaps. Gaps in an iteration may be completely closed subject to the following condition [Aik88]:

Condition 7.1. *Let iterations i and $i+c$ be alike. Assume that all iterations i through $i+c$ have the same number of maximal regions. Let A_j^k be the j-th maximal region on the k-th iteration, where $i \leq k \leq i+c$. If*

$$gap(A_j^k, A_{j+1}^k) \leq gap(A_j^{k+1}, A_{j+1}^{k+1})$$

then the gaps in iteration $i+c$ and i may be completely closed.

Next, the OPT algorithm is presented.

Algorithm 7.2. The input to this algorithm is a singly nested loop $\mathcal{L}$ with a dependence graph $G(V, E)$. The output is an optimal schedule of $\mathcal{L}$. Let $\mathcal{L}_k$ represent the k-th iteration of $\mathcal{L}$.

kernelfound $\leftarrow$ *false*
$i \leftarrow 0$

while $\neg kernelfound$ **do**
  Unroll the loop body once
  $i \leftarrow i+1$

  Schedule operations in the unrolled loop body $G'(V', E')$ as soon as possible
  for each iteration $\mathcal{L}_k \in \mathcal{L}$, where $0 \leq k \leq i-1$ **do**
    ShrinkGaps ($\mathcal{L}_k$, $\mathcal{L}_i$)
  endfor

  // Look for a kernel, a sequence of steps holding one copy of each operation
  // of the loop body. This implementation is simple but very inefficient.
  Let s_i be the set of operations at control step i
  Let L be the length of the schedule

  // It is possible that multiple sequences of control steps could be valid kernels.
  // Search starting at the end of the schedule first to find the latest (and best) one.
  for $j = L..1$ **do**
    $S \leftarrow \emptyset$
    $i \leftarrow j$
    while $S \neq V \wedge i \leq L$ **do**
      if s_i contains multiple copies of an operation or $s_i \cap S \neq \emptyset$ **then**

```
            break // failure, exit while loop
        else
            S ← S ∪ s_i
        endif
        i ← i + 1
    endwhile

    if S = V then
        kernelfound ← true
        break // success, exit for loop
    endif
  endfor
endwhile
```

The function *ShrinkGaps* ($\mathcal{L}_k$, $\mathcal{L}_i$) shrinks the gaps in iteration L_i if the two iterations satisfy the conditions discussed earlier. In theory, there are loops for which the strategy of making iterations look alike cannot succeed in polynomial time if dependence cycles have distances greater than one, because the length of a kernel based on this approach is at least the least common multiple of the cycle distances. If there are many distinct dependence cycles with relatively prime distances, then the complexity of the algorithm is potentially exponential in the number of operations in the loop body. In practice, this behavior is never observed because almost all dependence cycles have distance one and almost all of the rest have distance two.

The efficiency of the algorithm is dependent on the cost of computing a greedy schedule. This can be easily done using a modified topological sort of the dependence graph. The cost is proportional to the number of operations scheduled.

Example 7.2. Consider the data dependence graph shown in Figure 7.5(a) originally proposed as an example in [Cyt84]. The corresponding greedy schedule (i.e., operations are scheduled as early as possible subject to data dependences) is shown in Figure 7.5(b). Even though the greedy schedule corresponds to maximum parallelization, however, from Figure 7.5(b) one observes that a kernel does not emerge as successive iterations are scheduled.

In Figure 7.5(a), the cycle containing the operations v_2, v_5, v_8 has the greatest slope (three). Operations v_3, v_4 and v_7 are dependent on this cycle and thus have the same slope. All other operations have a slope of zero (0/1). From Figure 7.5(b) one notes that the schedule is eventually split into two groups that repeat every iteration, one with a slope of three, the other with a slope of zero. In an attempt to generate a kernel, the algorithm reschedules the operations not on the critical path so that they have the same slope as the operations on the critical path. In Figure 7.5(b), operations with a slope of zero (0/1) in iterations four and five could be delayed without affecting the length of the schedule. Eliminating the gaps in the iterations produces the (optimal) schedule shown in Figure 7.6. The boxed area is the kernel

Iteration

1	2	3	4	5	6	7
v_1 v_2 v_3	v_1	v_1				
v_4 v_5 v_6 v_9	v_9	v_9				
v_7 v_8 v_{10} v_{11} v_{12}	v_3 v_{11} v_{12}	v_{11} v_{12}	v_1			
v_{13}	v_2 v_4 v_{13}	v_{13}	v_9			
v_{14}	v_5 v_6 v_7 v_{14}	v_6 v_{14}	v_{11} v_{12}			
v_{15} v_{16} v_{17}	v_{15} v_{16} v_{17}	v_3 v_{15} v_{16} v_{17}	v_{13}	v_1		
	v_8 v_{10}	v_4 v_{10}	v_6 v_{14}	v_9		
		v_2 v_7	v_{15} v_{16} v_{17}	v_{11} v_{12}		
		v_5	v_{10}	v_{13}	v_1	
		v_8	v_3	v_6 v_{14}	v_9	
The kernel			v_2 v_4	v_{15} v_{16} v_{17}	v_{11} v_{12}	
			v_5 v_7	v_{10}	v_{13}	v_1
			v_8	v_3	v_6 v_{14}	v_9
				v_2 v_4	v_{15} v_{16} v_{17}	v_{11} v_{12}
				v_5 v_7	v_{10}	v_{13}
				v_8	v_3	v_6 v_{14}
					v_2 v_4	v_{15} v_{16} v_{17}
					v_5 v_7	v_{10}
					v_8	v_3
						v_2 v_4
						v_5 v_7
						v_8

Figure 7.6 : Optimal Schedule corresponding to the dependence graph in Figure 7.5(a)

for the loop; scheduling additional iterations (without gaps) reproduces these three control steps. Note that no operation on the critical path has been delayed as a result of re-scheduling.

7.4 General Handling of Conditionals

The problems caused by conditional branches in non-loop code have been described in some detail in previous chapters. These problems are further exacerbated by the presence of loops. Indeed, none of the techniques described previously directly apply to the exploitation of parallelism beyond the boundaries of loop iterations. For example, basic block techniques are intrinsically (by definition) applicable only to non-loop code. Augmented basic block techniques and trace scheduling could well pick for compaction blocks connected through a back edge, but cannot accommodate multiple iterations without unrolling. Similarly, percolation scheduling could be applied methodically, repeatedly across back edges, and may yield something akin to one of the software pipelining schemes described later in this chapter [Ebc87], but without—and this is critical, as we will see—providing a guarantee of termination (also called *convergence* in the literature). In this section we discuss the main approaches to general software pipelining of loops with conditionals.

7.4.1 Perfect Pipelining

Perfect pipelining [AN87] is a general approach for software pipelining of loops with conditionals. Like OPT, the kernel recognized by perfect pipelining corresponds to unbounded unrolling and compaction. The primary difference is that perfect pipelining finds a kernel for all paths despite arbitrary flow of control within the loop body. As discussed in Section 7.1, compaction of operations during software pipelining must be constrained in some way to guarantee the formation of a kernel; for example, OPT constrains the compaction process by shrinking the scheduling gaps within an iteration. Next, we discuss the constraints on compaction of operations in the context of perfect pipelining.

Because perfect pipelining works with both unrolled loops and across multiple control paths, we need more notation than in previous chapters to describe how perfect pipelining works. We view a (possibly parallelized) loop as a set $\mathcal{N}$ of nodes $n_0, \ldots, n_{m-1}$, where each node $n \in \mathcal{N}$ contains a set of operations *ops*(n) that are executed in parallel. Execution always begins at a distinguished *start* node n_0 and proceeds sequentially from node to node. When control reaches a particular node, all operations in that node are executed in parallel; the assignments update the registers or locations in the memory and the conditionals return the next node in the execution sequence. Operations evaluated in parallel perform all reads before any write. Write conflicts within a node are not permitted (that is, two operations may not write to the same register or memory location).

Compaction

Perfect pipelining applies to a single input loop $\mathcal{L}$, and we superscript operations with the iteration they come from in $\mathcal{L}$. After scheduling, nodes may contain operations from different iterations; i.e., it is possible that $x^i, y^j \in$ *ops*(n) and $i \neq j$.

Let $u^i\mathcal{L}$ stand for loop $\mathcal{L}$ unrolled i times such that $u^i\mathcal{L}$ consists of unrolled iterations $\mathcal{L}_0, \ldots, \mathcal{L}_{i-1}$ and $Cu^i\mathcal{L}$ stand for a compaction operator C applied to $u^i\mathcal{L}$. For example, a greedy scheduler that just schedules operations as early as possible is a compaction operator. We now introduce a relation $\leq_p$ that compares two different schedules for the same operations.

Definition 7.3. *If $\mathcal{L} \leq_p \mathcal{L}'$, then operations in $\mathcal{L}'$ are executed at least as early as the corresponding operations in $\mathcal{L}$.*

For the perfect pipelining algorithm, a compaction operator C must satisfy the following constraints:

I) **Boundedness** : A compaction operator is *bounded* if there is a non-negative integer k for all nodes in $Cu^h\mathcal{L}$ such that if $v_1^i \in$ *ops*(n), then for any $j > i + k$ we have $v_2^j \notin$ *ops*(n).

II) **Locality**: The compaction operator itself is a program with a fixed-size memory—it cannot make use of an arbitrary size state to decide how to schedule operations.

III) **Monotonicity** : A compaction operator C should perform better with larger unrollings of the loop. Therefore, C must satisfy the following:

$$Cu^i \mathcal{L} \leq_p Cu^{i+1} \mathcal{L} \ \forall i$$

Boundedness says that the operations at any control step in the unrolled and compacted loop are drawn from at most k consecutive iterations of the original loop.[3] Boundedness guarantees that for any particular program there is a finite number of different combinations of operations that can be scheduled in a control step—thus, the scheduler is bound to repeat some combinations of operations if the loop is unrolled enough.

However, boundedness is only a necessary but not sufficient condition for convergence. Additionally, the compaction algorithm itself cannot use arbitrary amounts of internal state in deciding how to schedule operations. To see why, assume we are scheduling a program where it happens that there are only two legal combinations of operations A and B that can be scheduled in any node. Even though the scheduler must repeat A nodes and B nodes very frequently in the schedule, it is still possible to have non-repeating sequences of A and B nodes on a path, just as it is possible to write irrational numbers with only two digits. By restricting the compaction algorithm to have only a fixed size internal state, we guarantee that when in the same internal state and facing the same choices of operations to schedule the scheduler will always make the same decisions—i.e., it is guaranteed to fall into a repeating pattern on all execution paths for sufficiently large unrollings of a loop.

While boundedness and locality are sufficient to guarantee convergence, monotonicity simply says that the scheduler can only improve the schedule as it unrolls the loop more—that is, the scheduler shouldn't make the schedule worse by considering more iterations. The monotonicity constraint plays an important role in proving that perfect pipelining finds a schedule that is equivalent to $Cu^\infty \mathcal{L}$ and also better than $Cu^i \mathcal{L}$ for any i. However, the constraint is stronger than necessary. For example, in the case of bounded resources, a loop unrolled and compacted i times may be faster than a loop unrolled and compacted $i + j$ times because the smaller loop fits the resources

[3] This definition of boundedness is stronger and also simpler than the original condition in [Aik88]

better. A relaxed form of the monotonicity condition that works for bounded resources is discussed in [Aik88].

The Algorithm

Perfect pipelining is parameterized by a compaction algorithm; any compaction algorithm satisfying the boundedness, locality and monotonicity constraints is guaranteed to converge and produce a result equivalent to full unrolling and compaction.

We now focus on one family of compaction algorithms for use with perfect pipelining based on the core transformations of percolation scheduling [Nic85b]. The transformations, as discussed in Section 5.2, involve only adjacent nodes of the dependence graph.

The input is a single (unnested) loop $\mathcal{L}$. Let V denote the set of operations that form the loop body of $\mathcal{L}$. Operations are classified as *assignments* that read and write a register or a memory location, *conditionals* that affect the flow of control, and a distinguished operation *stop*.

We define the following additional functions on the nodes $\mathcal{N}$:

- *branch*(n) is the set of all conditional branches in n,
- *succ*(n) is the set of all successor nodes of n, and
- *succ-on-branch*(n, p_i) maps a node n and a path through the dag of conditional branches in n to a successor node $n' \in succ(n)$.

We write $n \rightarrow_p n'$ as a shorthand for $\textit{succ-on-branch}(n, p) = n'$.

The following notation for talking about the same set of operations from different iterations is quite useful in explaining the perfect pipelining algorithm.

Definition 7.4. *Given a set of operations O, let O^k be the same set of operations from k iterations later; i.e., $x^i \in O$ if and only if $x^{i+k} \in O^k$.*

We now define what it means for two nodes to be equivalent in a schedule.

Definition 7.5. *Two nodes n_1 and n_1' are* equivalent *in $Cu^i\mathcal{L}$ if, for any k there is an unrolling j such that in $Cu^{i+j}\mathcal{L}$ there is a path of length k beginning at n_1*

$$n_1 \rightarrow_{p_1} n_2 \rightarrow_{p_2} \cdots \rightarrow_{p_k} n_k$$

if and only if there is a path beginning at n'_1

$$n'_1 \rightarrow_{p_1} n'_2 \rightarrow_{p_2} \cdots \rightarrow_{p_k} n'_k$$

such that there is a constant c *where for all nodes* n_i *we have* $ops(n_i)^c = ops(n'_i)$.

Intuitively, two nodes n and n' are equivalent if they are indistinguishable in the fully unrolled and compacted loop $Cu^{\infty}\mathcal{L}$—i.e., starting at either node executes the same sequence of operations offset by c iterations.

Example 7.3. For simplicity we illustrate the idea of equivalent nodes using an example without conditionals. Consider the loop shown in Figure 7.7(a). The corresponding data dependence graph is shown in Figure 7.7(b). The latencies of the operations are shown in Figure 7.7(b).

In Figure 7.7(d), although control steps 2 and 6 contain the same set of operations they are not equivalent. In particular, note that the set of operations in the successor control steps 3 and 7 are not the same. In contrast, the nodes corresponding to control steps 6 and 8 are equivalent, which is evident from the fact that their successors

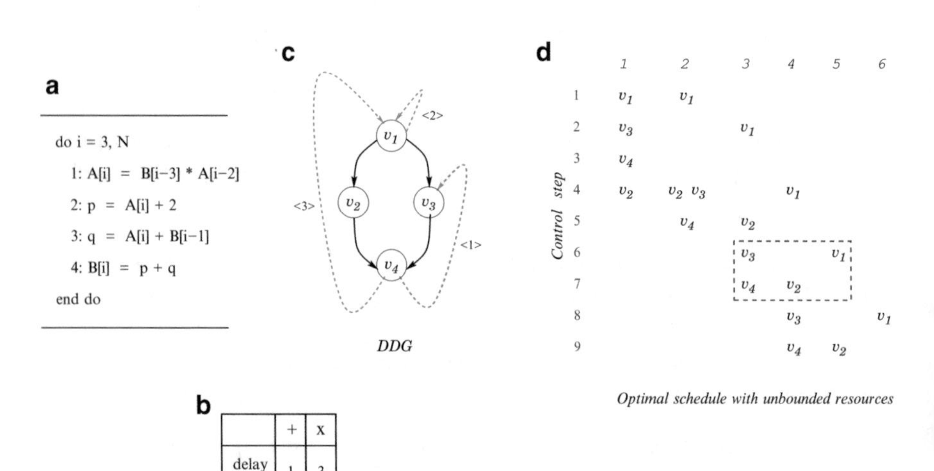

Figure 7.7 : Equivalence of nodes

(and successors of successors and so on) keep repeating during the rest of the schedule. The box around steps 6 and 7 shows the kernel.

The specific compaction strategy we present for perfect pipelining generates a schedule in an incremental fashion. Initially, the schedule is empty and some set of operations $\mathcal{A}$ are available for scheduling (have no dependences on other operations). The algorithm chooses a subset of $\mathcal{A}$ as the start node, and then recursively schedules the successors of the start node by choosing a subset of the (updated) available operations for each successor node, and so on.

Algorithm 7.3. Perfect Pipelining
The input to the algorithm is an initial set of available operations $\mathcal{A}$ and a window size k. The output is a software pipelined schedule.

Seen $\leftarrow \emptyset$ // set of previously seen states
$\mathcal{S} \leftarrow \{\langle n, \mathcal{A}\rangle\}$ // n is an empty node

while $\mathcal{S} \neq \emptyset$ **do**
  choose $\langle n, \mathcal{A}\rangle \in \mathcal{S}$
  $\mathcal{S} \leftarrow \mathcal{S} - \{\langle n, \mathcal{A}\rangle\}$

  // *schedule-node* schedules available operations from $\mathcal{A}$ in n.
  schedule-node$(n, \mathcal{A})$

  // *next* returns each successor n_i of n together with p_i such
  // that $n \rightarrow_{p_i} n_i$ and the available operations $\mathcal{A}_i$ at n_i.
  $\{\ldots, \langle p_i, n_i, \mathcal{A}_i\rangle, \ldots\} =$ *next* (n)
  for each $\langle p_i, n_i, \mathcal{A}_i\rangle$ **do**
    if $\langle n', \mathcal{A}_i^c\rangle \in$ **Seen** (for some n' and integer c) **then**

      // If we have seen the same set of available operations $\mathcal{A}_i$ before, then n_i
      // is a repeated state. Replace successor of n with previously scheduled node n'.
      succ-on-branch$(n, p_i) \leftarrow n'$
    else

      // Otherwise, add $\langle n_i, \mathcal{A}_i\rangle$ to the set of previously seen
      // states **Seen** and the worklist $\mathcal{S}$.
      Seen $\leftarrow$ **Seen** $\cup \{\langle n_i, \mathcal{A}_i\rangle\}$

$\mathcal{S} \leftarrow \mathcal{S} \cup \{\langle n_i, \mathcal{A}_i \rangle\}$
endif
endfor
endwhile

Algorithm 7.3 schedules operations in node n by invoking the procedure *schedule-node*, which in turn calls the procedure *schedule* to repeatedly select one operation from $\mathcal{A}$ to schedule in n (see below). The operations in $\mathcal{A}$ and the function *schedule* are subject to the following constraints:

- If $v^i \in \mathcal{A}$, then $x^{i+k} \notin \mathcal{A}$ for every x. In other words, operations in $\mathcal{A}$ span a window of at most k consecutive iterations.
- *schedule* is a function of two arguments, a the set n of operations already scheduled in the current node and the set of available operations $\mathcal{A}$. Furthermore, if $x^i = \textit{schedule}(n, \mathcal{A})$ then $x^{i+k} = \textit{schedule}(n^k, \mathcal{A}^k)$.

The first requirement ensures that the algorithm is bounded. The second requirement guarantees that scheduling choices are a (pure) function of a fixed amount of state. In particular, the choice of which operation to schedule can depend only on the number of iterations between operations (which is bounded by the window size k) but not on the actual value of the iterations of operations (which is not bounded).

Algorithm 7.4. *schedule-node* $(n, \mathcal{A})$

// Operations are scheduled iteratively because it is possible
// that two operations are both available but cannot be scheduled together.
$\textit{ops}(n) \leftarrow \emptyset$
$v \leftarrow \textit{schedule}(n, \mathcal{A})$
while $v \neq \textit{null}$ **do**

// *update* moves the operation v to n and returns n and
// an updated set of available operations
$n, \mathcal{A} \leftarrow \textit{update}(n, \mathcal{A}, v)$
$v \leftarrow \textit{schedule}(n, \mathcal{A})$
endwhile

The algorithm determines the set of available operations for each node $n_i \in \textit{succ}(n)$ using Algorithm 7.5 (described below). If at any point the algorithm encounters the same set of available operations a second time, it uses the previously scheduled (equivalent) node. Note that it is not necessary for the equivalent nodes to be from the same path. The constraints, discussed on page 186, ensure that having the same set of available operations for two nodes implies that all possible branches from the two nodes are also the same, a result which is proven in [AN95].

The check whether $\mathcal{A} \in \mathbf{Seen}$ can be implemented efficiently in the following way. For a window size of k iterations, operation availability for iterations j through $j + k - 1$ is represented as a bit vector of length kN, where N is the number of operations in the sequential loop. The bit $hN + i$ is 1 if operation $v_i^{j+h} \in \mathcal{A}$, otherwise it is 0. When iteration j is completely scheduled (i.e., when the first n bits are all 0) the bit vector is shifted left n bits, discarding information for iteration j, and the last n bits are set to reflect the availability of operations in iteration $j + k$. With this representation, checking whether the same availability information has been seen before only requires checking whether the same bit vector has been seen before, which can be implemented very efficiently through hashing.

Algorithm 7.3 terminates when the work list $\mathcal{S}$ is empty. The set $\mathcal{S}$ decreases in size whenever the current set of available operations $\mathcal{A}$ has an equivalent set $\mathcal{A}^c$ in $\mathbf{Seen}$. As discussed above, since each set of available operations is drawn from at most k consecutive iterations and, furthermore, only the relative distance between iterations matters and not the actual value of the iterations of the operations, every possible set of available operations can be represented as a bit vector of length kN, and therefore there are only 2^{kN} distinct possible sets of available operations. Inspection of the algorithm shows that the worklist S only grows when the set $\mathbf{Seen}$ of previously seen sets of available operations also grows, and furthermore nothing is ever removed from $\mathbf{Seen}$. Therefore $\mathbf{Seen}$ must eventually reach a point where no more sets of available operations can be added, after which nothing more will be added to S and the algorithm will terminate.

We now discuss in detail how to compute the set of operations available for scheduling at a node. Let *write*(v) (resp. *read*(v)) be a set of all the locations [4] an operation v may write (resp. *read*(v)). The set *kill*(v) consists of locations v always writes to. Two different sets *write*(v) and *kill*(v) are defined because dependence analysis

[4] A location is either a memory location or a register.

must be conservative in general; as it is not always possible to determine at compile time exactly which locations an operation may read or write. An operation v may depend on another operation u if :

$$\begin{aligned} depends(u, v) \quad \equiv \quad & (write(u) \cap (read(v) \cup write(v)) \neq \emptyset) \vee \\ & (write(v) \cap (read(u) \cup write(u)) \neq \emptyset) \end{aligned}$$

We extend *depends* to work on sets of operations as follows:

$$depends(O, v) \equiv \bigvee_{o \in O} depends(o, v)$$

The set of operations $nodeps(n)$ that can be potentially moved to a node n, i.e. operations that are not dependent on an intermediate operation, is given by:

$$nodeps(n) = ops(n) \cup ((\bigcup_{n' \in succ(n)} nodeps(n')) - \{v | depends(ops(n), v)\})$$

The set $nodeps(n)$ can be computed, for all n, by a single bottom-up traversal of the control flow graph of a completely unrolled loop. However, not all the operations in $nodeps(n)$ can be scheduled in node n as any operation that kills a live reference in node n cannot be scheduled in n. For example, in Figure 7.8 operation v_2 is not available for scheduling at the first node, because it will kill the value of j read by operation v_3.

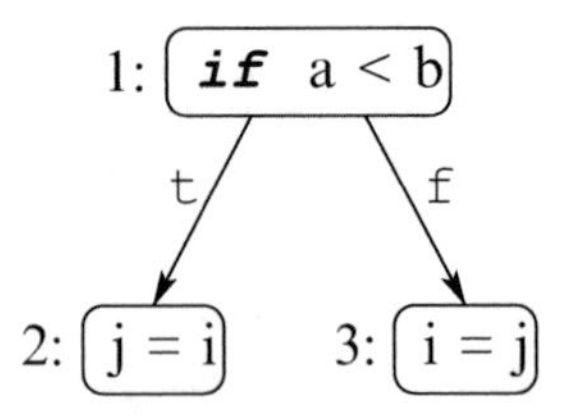

Figure 7.8 : Operation v_2 can kill a reference live at v_1

A reference to a location l is live at a node n if there is a node reachable from n where l is (potentially) read and there is no intervening write to l. While moving an operation v in the schedule, it is necessary to check if v kills any of the live references in its new position. However, in deciding whether or not v will kill live references in its new position one should not count references of v itself in its current position. The set of locations live at a node n modulo operation v is given by :

$$live(n,v) = read(ops(n) - \{v\}) \cup ((\bigcup_{n' \in succ(n)} Z_{n'}) - kill(ops(n) - \{v\})$$

$$\text{where } Z_{n'} = \begin{cases} live(n',v) & \text{if } v \in nodeps(n') \\ live(n',stop) & \text{otherwise} \end{cases}$$

The two cases in the definition of $Z_{n'}$ distinguish between the cases where occurrences of v can or cannot be blocked by dependences. If there is an occurrence of v on a path that is not blocked by dependences (i.e., $v \in nodeps(n')$) then that occurrence of v is discounted in the live reference computation (i.e., $live(n', v)$). If there is no occurrence of v that can potentially move, then all live references are counted (i.e., $live(n', \text{stop})$ counts all references, since stop has no effect on the store). As with the computation of *nodeps*, $live(n, v)$ can be computed for all states n and operations v by a single bottom-up traversal of the control-flow graph.

Recall that in any node operations can be available from at most k consecutive iterations. Let i_{min} be the minimum iteration number of any operation in $\mathcal{A}$. Then the set of available operations at node n is:

$$avail(n) = nodeps(n) - \{v^i | (write(v^i) \cap live(n, v^i) \neq \emptyset) \wedge (i - i_{min} < k)\}$$

Algorithm 7.5 inserts a new, empty node n_i for each *succ-on-branch*(n, p_i) on branch $p_i \in branch(n)$ as well as the set of operations available $\mathcal{A}_i$ for scheduling.

Algorithm 7.5. $next(n)$

for each $p_i \in branch(n)$ **do**
  let n_i be an empty node
  succ-on-branch$(n, p_i) \leftarrow n_i$
  $\mathcal{A}_i \leftarrow avail(n_i)$
endfor
return $\{\ldots, \langle n_i, \mathcal{A}_i \rangle, \ldots\}$

Algorithm 7.6. *update* $(n, \mathcal{A}, v)$

if v is an assignment **then**
  $\mathcal{A} \leftarrow \text{unify}(n, \mathcal{A}, v)$

else
  $\mathcal{A} \leftarrow$ move-test$(n, \mathcal{A}, v)$
endif
Let $Y \leftarrow \{n_k | \exists$ a path $\langle n, \ldots, n_k \rangle$ s.t. $\forall i,\ 1 \leq i \leq k$ and $v \in \mathit{nodeps}(n_i)\}$
update *nodeps* and *live* for $n \in Y$ and any newly added nodes
return $n, \mathcal{A}$

Algorithm 7.6 moves v from its original place in the sequential schedule to n. However, it is possible that some copies of v cannot be moved on some other path(s). When v is a conditional, the control flow of $\mathcal{L}$ must be modified in order to ensure correctness. The procedure also updates the sets *nodeps* and *live* where necessary. Since both *nodeps* and *live* are computed bottom-up, the sets can be updated in a single bottom-up pass over those paths where the information may have changed.

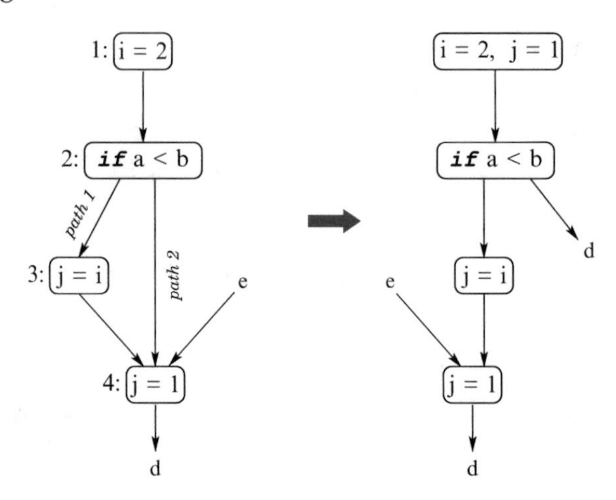

Figure 7.9 : Moving an operation in presence of dependences

Moving an operation v while preserving program semantics is a little subtle because v may be available at n on one path but may be blocked by data dependences on some other path(s). In addition, there may even be paths from other nodes in $\mathcal{S}$ (the work list) to v. For example, in Figure 7.9 operation v_4 is available at the node 1 along path 2 as v_4 does not kill any references live at node n_1 and is not dependent on any operation along path 2. However, along path 1, v_4 is dependent on operation v_3; clearly, v_4 cannot be deleted from path 2. In addition, there is a path, represented by the incoming edge from e, from other nodes to v_4.

The percolation scheduling transformations automatically ensure that operations are preserved on the paths where the operation is not being moved. When an operation is an assignment, the operation can be moved using the *move-op* transformation of percolation scheduling, as discussed in Section 5.2. Similarly, a conditional operation can be moved using the *move-test* transformation of percolation scheduling. The sets *nodeps* and *live* are updated for all those nodes lying on the path along which the operation v is moved up. Nodes that are duplicated during the transformations retain the *nodeps* and *live* information.

For clarity we have presented a version of perfect pipelining without resource constraints, multi-cycle operations and nested loops. The algorithm is easily extended to handle these features. For a general resource model, the algorithm can be modified by associating a reservation table describing the resource usage with a node n [DSTP75, TTTT77, Bae80]. The algorithm can be extended for multi-cycle operations by modeling an i-cycle operation v as a chain of i single-cycle operations; operations that depend on v are dependent on the last operation in the chain. To guarantee a legal schedule, all the operations of a chain must be executed in successive cycles without interruption. Multi-level nested loops can be handled by unrolling multiple loops.

Example 7.4. Consider the loop shown in Figure 7.10(a). The corresponding control flow graph is shown in Figure 7.10(b). In Figure 7.10(b), operation v_6 corresponds to the bounds check and index increment `if i++ < N`. Initially, the set of available operations at the start node n_0 is $A_0 = \{v_1^1, v_2^1, v_6^1\}$. Since v_6 is a conditional, there are two successors of the first state, nodes n_3 and n_1 in Figure 7.10(c).[5] Observe that the set of available operations is different for the two successors. If operation v_6^1 in node n_0 evaluates to false, then the loop terminates. The set of available operations A_1 for node n_1 is $\{v_3^1, v_4^1, v_5^1\}$. Since write conflicts are not permitted, operations v_3^1 and v_4^1 cannot be scheduled together even though both are available in node n_1. Again, since v_3^1 is a conditional, node n_2 has two successors. For node n_2, the set of available operations is $\{v_5^1\}$.

[5]The node numbers are marked on the top left of each box.

a

```
do i = 1, N
   1: A[i] = f (A[i-1])
   2: j = i
   3: if A[j] < 0
   4:     B[j] = A[j]
      else
   5:     B[j] = -A[j]
end do
```

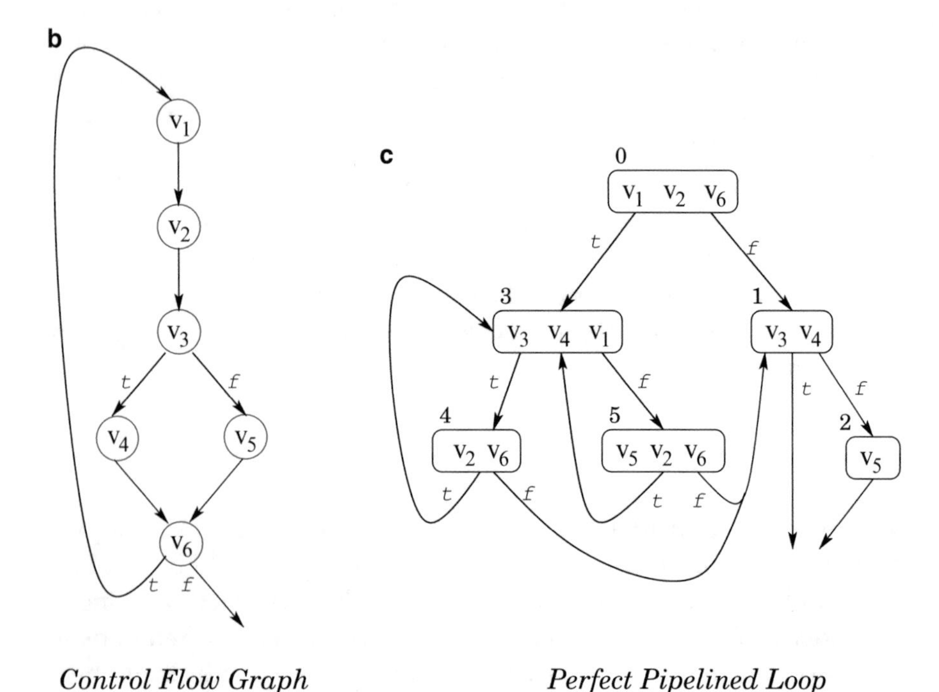

Figure 7.10 : Perfect Pipelining with multiple paths

If v_6^0 evaluates to true the set of available operations A_3 for node n_3 is $\{v_3^1, v_4^1, v_5^1, v_1^2\}$. Note that the operation v_1^2 from the second iteration is available for scheduling in parallel with statements from the first iteration. A subtle point is that operation v_2^2 is not available for scheduling, even though all reads take place before all writes and all operations from the first iteration that read variable j are available in A_3. Operation v_2^2 is not available because both the operations v_4^1 and v_5^1 cannot be scheduled in node n_3. Even though all operations that read j are in A_3, not all of these can be scheduled in n_3, and this fact prevents operations that write j from being available.[6]

Assume that the operations v_3^1, v_4^1 and v_1^2 are scheduled in node n_1. Since operation v_3^1 is a conditional, there are two successors of n_1. For node n_4, the set A_4 is $\{v_2^2, v_6^2\}$. The set of available operations for the successor of n_4 on the path where v_6^2 is true is $\{v_3^2, v_4^2, v_5^2, v_1^3\}$. Note that, except for the superscripts, this set is exactly the same as A_3. The superscripts are just a way of keeping track of the iteration of each operation; the sets have the same operations. Rather than continue scheduling at this point, the algorithm simply makes n_3 a successor of n_4. Similarly, the set of available operations for the successor of n_4 when v_6^2 evaluates to false is $\{v_3^2, v_4^2, v_5^2\}$. Except for superscripts, this is exactly the same as A_1. As before, the algorithm makes n_1 a successor of n_4.

Let us now consider the successor n_5 of n_3 when v_3^1 evaluates to false. The set of available operations for node n_5 is $\{v_5^1, v_2^2, v_6^2\}$. The sets of available operations for the two successors of n_5 are the same as for n_4 and scheduling proceeds just as it did for n_4. The algorithm terminates with the schedule in Figure 7.10(c).

7.4.2 Enhanced Pipeline-Percolation Scheduling

Enhanced Pipeline-Percolation Scheduling (EPPS) [EN90], based on an earlier algorithm *pipeline scheduling* [Ebc87], integrates code motion with scheduling (like perfect pipelining) for software pipelining a loop. However, unlike the kernel recognition techniques discussed so far, EPPS does not need to search for a kernel in the unrolled code. At every step EPPS maintains a valid loop (so in principle the algorithm could halt at any point with a correct program) and uses a special mechanism to detect when no further improvement can be made.

[6]The careful reader may have noticed that the algorithm we have given for computing the available operations actually doesn't take into account that write operations can be moved into nodes that already read from the same location—i.e., the presented algorithm treats operations involved in write-after-read dependences as unavailable. The modifications needed to support this case are tedious but straightforward.

One of the sources of variation in software pipelining algorithms is the different assumptions about the capabilities of the underlying hardware. While the ideas in EPPS could easily be adapted to target other architectures, the published version of EPPS schedules code for a machine that executes *tree* instructions. Despite its name, a tree instruction is modeled as a dag. Conditionals and assignments may be placed anywhere in the dag except at terminal nodes, which are exit points from the tree instruction and name the next tree instruction to be executed. Execution of a tree instruction (conceptually) involves two phases:

i) *Path-selection phase* - A unique path from the root to a leaf of the tree instruction is selected. The selection is done based on the old values of condition code registers that were set by previous tree instructions. If a conditional is true, then the left branch is taken, otherwise the right branch is taken.

ii) *Execution phase* - The arithmetic and memory operations along the selected path are executed in parallel. All operations read the old values of register and memory locations (i.e., those written by previous tree instructions). If more than one operation writes to a location l then the operation closest to the leaf determines the final value in l. The results of the operations are written into the register file/memory and control jumps to the next tree instruction pointed to by the terminal node.

EPPS alternates two kinds of transformations to parallelize the loop. In the first step, it takes the instruction(s) at the top of the loop (the entry points to the loop) and moves as many operations into these instructions as possible while respecting resource constraints. To a first approximation, these moves are done using the migrate transformation (Section 5.4.1) built on top of percolation scheduling's core transformations, but any heuristic could be used to move operations into the top instruction(s). When no more operations can be moved into one of the top instructions (either because of dependence or resource constraints) the instruction is said to be *filled*. Importantly, when these moves are done the backedge(s) of the loop are ignored— the moves are done only on the acyclic loop body, and only the instructions with no predecessors in the loop body are filled.

Once the instructions at loop body entry points are filled, the second step is to *rotate* the loop. Consider an instruction n which is an entry point into the loop. Some of the predecessor edges of n are backedges in the control-flow graph and some predecessor edges come from outside of the loop. Instruction n is duplicated; one copy is made the target of the backedges and the other copy is made the target of the non-backedges. Then the copy of n that is the target of the loop backedges is moved to the end of the loop and n's successors are labeled as the entry points to the loop. Loop rotation is most easily illustrated with an example. Consider the following sequential loop:

$x \leftarrow 1$
$y \leftarrow 1$
L: **if** $x > n$ **goto** M
$x \leftarrow x + y$
$y \leftarrow y + 1$
goto L
M: ...

To rotate this loop we want to move the test with label L to the end of the loop body. Since L has two control-flow predecessors, one of which is a backedge of the loop and one which is not, we duplicate the instruction at L and place one copy after each of its predecessors. We can now write the loop so that the successor of L is the entry point to the loop; below we have introduced a new label L' to name this instruction and make it the new target of the backedge of the loop:

$x \leftarrow 1$
$y \leftarrow 1$
if $x > n$ **goto** M
L′: $x \leftarrow x + y$
$y \leftarrow y + 1$
if $x > n$ **goto** M
goto L′
M: ...

Once the loop has been rotated, EPPS repeats step 1 on the new entry point(s) of the loop. Note that after rotation, the filled instruction that is moved to the bottom of the loop now represents the first instruction of the *second* iteration of the loop before rotation. Thus, by moving operations in this instruction up using compaction operators multiple iterations of the original loop are overlapped and software pipelining occurs.

This process of filling the top nodes of the loop body and then rotating them to the bottom of the loop body is iterated until no unfilled nodes remain in the loop body. To guarantee termination, it is necessary for EPPS to place some restrictions on code motion; otherwise it is possible that the algorithm will fall into oscillating behavior, perhaps cycling through a number of distinct schedules over and over again. Two rules ensure the algorithm halts:

- Instructions are only filled once—an instruction is never considered for filling a second time, even if it again becomes a top node of the loop body. Thus, nothing is ever added to a filled instruction.
- The operations in filled instructions can be moved out of the instruction and in to another instruction that is being filled only if *all* of the operations in the filled instruction can move together. Thus, operations are only removed from a filled instruction if it will become empty as a result and can be deleted.

Even with these restrictions, it is still not immediate that EPPS terminates, because the core transformations of percolation scheduling introduce new unfilled instructions into the program when unfilled instructions are duplicated as part of compaction.

The argument that EPPS does in fact terminate is similar to, but much simpler than, the proof that trace scheduling terminates presented in Chapter 4. Let $p(n)$ be the number of paths that an instruction n occurs on in the acyclic loop body. We define a function $F(n)$ as follows:

$$F(n) = \begin{cases} p(n) & \text{if } n \text{ is unfilled} \\ 0 & \text{if } n \text{ is filled} \end{cases}$$

Theorem 7.1. *The quantity $\Sigma_n F(n)$ strictly decreases on each iteration of EPPS scheduling.*

PROOF. There are four interesting cases.

- Consider a core percolation scheduling transformation that duplicates an instruction i into instructions i' and i''. If i is an unfilled instruction, then by construction $F(i) = F(i') + F(i'')$, since percolation scheduling only duplicates instructions to preserve them on the same set of paths. If a filled instruction is duplicated, both of the copies are also marked as filled, and so $F(i) = F(i') = F(i'') = 0$. In both cases the sum $\Sigma_n F(n)$ is the same after the transformation as before.

- Loop rotation only duplicates filled instructions and so cannot increase $\Sigma_n F(n)$.
- If an instruction n is deleted because it becomes empty, then if n is filled $\Sigma_n F(n)$ remains unchanged and if n is unfilled $\Sigma_n F(n)$ decreases by $p(n)$.
- Each iteration of EPPS fills at least one unfilled instruction, which decreases $\Sigma_n F(n)$.

The first three cases preserve the value of $\Sigma_n F(n)$ or decrease it, while the last guarantees that the sum decreases by at least one on each iteration of EPPS. ■

When $\Sigma_n F(n) = 0$ there are no unfilled instructions remaining; it follows from Theorem 7.1 that EPPS always terminates.

We now present pseudo-code for the EPPS algorithm. We assume there are two functions *fill(n)* and *rotate(n)* implementing the filling and rotation operations described above.

Algorithm 7.7. Enhanced Pipeline-Percolation Scheduling
The input to the algorithm is a loop body represented as a control-flow graph $G = (V, E)$.

for each $v \in V$ **do**
  filled$(v) \leftarrow$ false
endfor

$V' \leftarrow \{v | v \in V$ and v is unfilled and has no predecessors in $G\}$
while $V' \neq \emptyset$ **do**
  // Move operations to the instructions at the loop entry points
  // subject to dependences and resource constraints
  for each $v \in V$ **do**
    fill(v) // updates nodes and edges of G
    filled$(v) \leftarrow$ true
  endfor
  for each $v \in V$ **do**

rotate(v) // updates nodes and edges of G
endfor
// find the new entry points in the updated control-flow graph
$V' \leftarrow \{v|v \in V$ and v is unfilled and has no predecessors in $G\}$
endwhile

Note that if an instruction with two successors within the loop body is rotated to the bottom of the loop both successors become loop entry points in the rotated loop. For this reason Algorithm 7.7 must deal with multiple loop entry points, because even if the original loop has a single entry point there will be multiple entry points after rotating the first instruction with multiple successors in the loop body. We now present a simple example of the application of EPPS.

Example 7.5. Consider the control flow graph and the corresponding data dependence graph shown in Figures 7.11(a) and 7.11(b).

Figure 7.11(c) shows the various stages of software pipelining the loop using Algorithm 7.7. The boxed instructions show the loop body immediately after the top node of the loop body is filled and immediately before loop rotation. In the first call to *fill*, data dependences prevent any compaction from taking place. After v_1 is rotated to the bottom, in the second call to *fill*, v_1 is moved up and scheduled together with v_2 (note that v_1 is in a filled instruction at this point, but

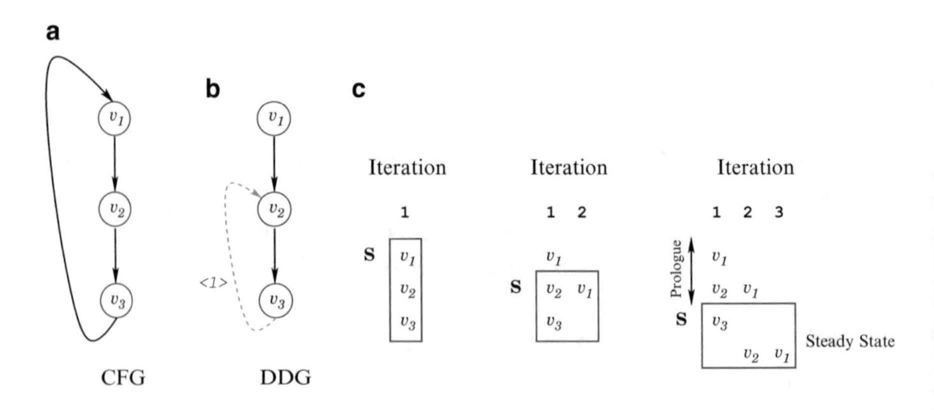

Figure 7.11 : Enhanced Pipeline-Percolation Scheduling

all the operations of that instruction, namely just v_1, can move up). The instruction with v_1 and v_2 is then rotated to the bottom but the loop-carried dependence between v_2 and v_3 and the rule that operations in filled nodes can only move together prevent any further compaction in the third call to *fill*. At this point both instructions in the loop body are filled and the algorithm terminates. The loop control and epilogue code is omitted for simplicity.

In the context of EPPS, the *move-op* transformation is enhanced in two ways. First, it is combined with the *unify* transformation so that as many copies of an assignment on different paths in a tree instruction are moved up as possible. Second, the targets of assignments can be renamed when being moved into a node to avoid overwriting values that are still live, and a copy operation is left behind in the original instruction to move the renamed value to the destination location (it is hoped that a subsequent register allocation phase will succeed in removing the copy by assigning both the source and target to the same register). Other enhancements include memory disambiguation for moving loads and stores, and a "combining" feature [NE89] (also known as *forward substitution*) for removal of flow dependences between seemingly dependent operations. The reader is referred to [EN90] for the details.

In the formulation of EPPS presented here, the prologue and epilogue code are generated automatically as part of the core percolation scheduling transformations and loop rotation. The original presentation is somewhat more involved, requiring the prologue and epilogue code to be constructed separately after the algorithm terminates [EN90].

7.4.3 Optimal Software Pipelining with Control Flow

Several optimal scheduling algorithms have been proposed for different classes of programs. For acyclic straight line programs with unbounded resources, list scheduling [Hu61, ACD74, Cof76] produces a time optimal schedule. It is well-known that the problem becomes NP-complete in the presence of resource constraints [GJ79]. In the presence of conditionals and unbounded resources, time optimal algorithms for acyclic programs are exponential in the number of operations [Fis84]; in the case of finite resources, time optimality is not well defined because of possible resource conflicts between operations executed speculatively but belonging to different paths.

For loops without conditionals, Aiken and Nicolau devised a time optimal software pipelining algorithm for parallelizing innermost loops, as discussed in Section 7.3. However, not all loops can be transformed to a corresponding semantically and algorithmically (SAE) equivalent time optimal loop [GSE89], for example, a `DOALL` or vectorizable loop [LB80, PKL80]. Introducing some artificial dependence constraints (e.g., at most one loop iteration may be initiated every control step) makes such loops amenable to scheduling with unbounded resources [Ebc88]. However, even in the presence of the such constraints, Gasperoni et al. [GSE89] proved that there exist loops for which a corresponding SAE time optimal loop does not exist.

In [Uht88], Uht proved that the resource requirement for optimal execution of loops with conditionals varies from exponential to polynomial in the length of the longest dependence cycle. Schwiegelshohn et al. [SGE91] proved that for a certain class of loops with conditionals there does not exist a corresponding SAE equivalent time optimal loop. Yun et al. [YKM01] generalized the above theoretical result and derived a necessary condition (discussed below) for the existence of time optimality of loops with conditionals.

7.5 Nested Loops

Several techniques have been proposed to extract fine-grain parallelism from nested loops. Loop quantization [Nic87] is a technique for unrolling multiple nested loops to facilitate extraction of parallelism that may be present in outer loops and not in the inner loop, or even across several nested loops. Once multiple nested loops have been unrolled, any technique for extracting ILP may be applied to the resulting loop body. In [AN90a], Aiken and Nicolau note that loop quantization followed by compaction is related to, but not the same as, the wavefront method for extracting iteration-level parallelism from nested loops [Mur71, Lam74, Kuh80, Wol82, Wol86].

The techniques discussed in the previous sections for software pipelining can be extended for exploiting ILP in nested loops. For example, GURPR extracts ILP from nested loops using a form of hierarchical reduction. First, the innermost loop is software pipelined via URPR (recall Section 7.2 on page 170). The kernel of the software pipelined innermost loop is then replaced with a single representative operation and the operations in the prologue, epilogue and the kernel operation are spliced into the next outer loop and the resulting loop body is software pipelined. If there is a conditional exit in the kernel, the kernel operation is a branch operation; branches may also exist in the prologue and epilogue. Note that the pipelining of the inner loop may give rise to multiple branches in the outer loop body.

Kim and Nicolau [KN92] extended the approach in [AN87] to exploit ILP in N-dimensional nested loops that do not contain conditionals. A different initiation interval is determined for each loop subject to the loop carried dependences, to exploit parallelism beyond the loop boundaries. The reader is referred to [KN92] for the details.

7.6 Procedure Calls

One situation we have not considered up to this point is the effect of a procedure call instruction within a loop. It may, of course, be desirable to be able to move operations around a procedure call, but because the executed procedure may have many more side effects than the call instruction itself it is important to account for dependences between operations within the loop and the called procedure. If interprocedural dependence information is available then nothing special need be done to use any of the scheduling algorithms we have presented. The usual situation, however, is that only intraprocedural dependence information is computed and call instructions must be handled as a special case.

Consider the example in Figure 7.12. Assume that v_1 writes a location read by the procedure and v_5 reads a location written by the procedure. It is incorrect for the read operation in the procedure of the i-th iteration to access the value written by v_1 in the $(i+1)$-th iteration. Similarly, it is incorrect for v_5 in iteration $i-1$ to read the value written by the procedure in the i-th loop iteration. In the presence of procedure calls, correctness can be preserved as follows:

- If the locations written by a loop operation v occurring before the call may be read by the procedure, or if the locations read by v may be written in the procedure, then v cannot move above the call of the previous iteration or below the call of the same

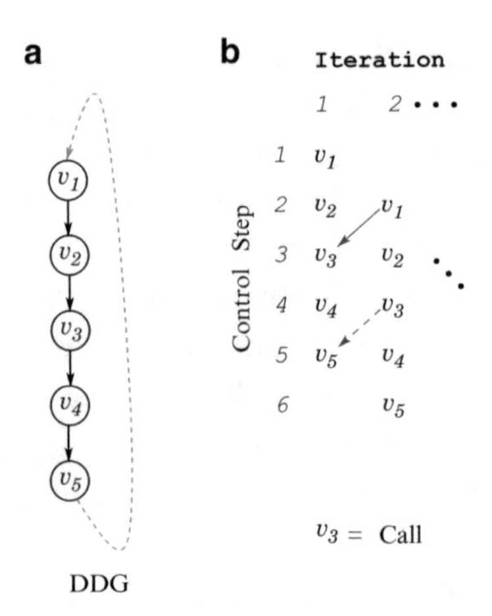

Figure 7.12 : Effect of Procedure Calls

iteration. We add a dependence from v_1 to the call within the iteration and a loop-carried dependence from the call to v with a distance of one. For example, in Figure 7.12 operation v_1 of the second iteration must be scheduled after operation v_3 of the first iteration. The dependence violation in the figure is shown by the bold arrow.

- If the locations read by a loop operation v occurring after the call may be written by the procedure, or if the locations written by v may be read by the procedure, then v must not be moved before the call in the current iteration nor after the call in the subsequent iteration. We add a dependence from the call to v and a loop-carried dependence from v to the call with distance one. For example, in Figure 7.12 operation v_3 of the second iteration must be scheduled after operation v_5 of the first iteration. The dependence violation in the figure is shown by the dashed arrow.

- Operations that are neither before nor after the call (e.g., in other branches of conditionals) are handled similarly; what is essential is that there be some conservative estimate of the read and write sets of the procedure so that dependences with other loop operations can be computed.

FURTHER READING

Many algorithms have been proposed for software pipelining. Some of the techniques proposed subsequent to the ones presented in this chapter include [EN89, Aik90, AN90b, Jai91, ME92, AN95, SC95, AGR95, GAG96, NN97]. Sánchez and Cortadella proposed a loop transformation-based software pipelining technique in [SC93, SC94]. Similarly, a significant amount of work has been done in the domain of register-sensitive software pipelining; some of the early work in this area includes [RLTS92, Huf93, NG93, WE93]. A quantitative evaluation of register pressure on software pipelined loops is available in [LAV98]. A throughput-oriented software pipelining approach is proposed by Sánchez and Cortadella in [SC96].

Software pipelining of nested loops was first discussed by Kim and Nicolau [KN92]; other works in this domain include [Ram94, Pet01]. A comparison of code generation techniques for software pipelined loops is presented in [RGSL96].

8

EPILOGUE

This book focuses on compiler-managed instruction level parallelism. While a great deal of the work on this topic has been only touched upon or mentioned only in references, the book does cover all of the major themes in the literature and in practice. However, the static compilation of instruction-level parallelism does not exist in isolation—there are related areas with close connections.

Most closely connected to compilation for ILP is dynamic (in contrast to static) instruction scheduling done in hardware and in particular Tomasulo's algorithm and its descendants [Tom67]. As discussed in Section 2.3, as early as the 1960's IBM's mainframe computers scheduled instructions out of order in hardware. Today variations on these ideas are integral to many high-performance CPUs. During program execution the processor examines a sliding window of as yet unexecuted instructions and performs instruction scheduling. Many of the important concepts in dynamic hardware instruction scheduling are the same as in the static exploitation of ILP:

- The scheduler performs dependence analysis on the window of instructions to determine which instructions, if any, are ready to execute and which must wait for other instructions to complete first. The scheduler guarantees that the execution order of dependent instructions is preserved. Two instructions are dependent if they access the same register and one operation writes that register, or if they both perform a memory operation to the same address (either a load or store) and one is a store.

A. Aiken et al., *Instruction Level Parallelism*,
DOI 10.1007/978-1-4899-7797-7_8

- The scheduler renames registers where possible to break false dependences between instructions. For example, if an instruction reads a register r and the next instruction writes r, the value of the write may be put in a temporary location r' to allow the two instructions to execute in parallel. Subsequent instructions that require the new value of r automatically have their argument renamed to r'.

- Scheduling may cross basic block boundaries by guessing, or *speculating*, that a conditional branch will be either true or false before the conditional test is evaluated, which is similar to selecting traces across conditional jumps in trace scheduling. Results of speculatively executed instructions are buffered until the result of the conditional test is known. If the branch prediction turns out to be correct then execution simply proceeds. If the branch prediction is incorrect then all the speculatively executed instructions and their results are discarded and execution restarts with the other branch of the conditional.

The primary differences between dynamic instruction scheduling in hardware and static scheduling by a compiler are, first, that the hardware has a more limited view of the instruction stream—the instruction window size is typically relatively small, and certainly much smaller than what can be considered in an off-line compilation. Thus, even for out-of-order processors static instruction scheduling can improve performance because of the broader view of the code taken by a compiler.

On the other hand, there is information available at run time that is unavailable to a static compiler. For example, at run time we have the exact addresses being loaded from and stored to memory, making the dependence analysis of memory operations more precise than can be achieved at compile time. As another example, modern processors keep profiling information during execution to predict which side of a conditional branch will be taken based on the recent history of the direction taken for that branch, and experience has shown that most branches are quite predictable. While compilers can also do a good job predicting branches (mostly because it is a safe bet that the branch deciding whether to do another iteration of a loop or exit the loop will overwhelmingly be biased towards executing more iterations), dynamic branch prediction works even better.

While instruction scheduling and software pipelining were invented to take advantage of the very fine-grain parallelism available at the level of individual instructions, there has been a recent trend toward applying these ideas at coarser granularities in parallel programming as well. In these approaches, the unit of computation to be scheduled is an entire function instead of a single instruction; the unit of data is typically a buffer, array or other aggregate structure instead of a single memory word or register. The goal in this setting is to ensure that while a function is occupying the processor some other useful work is going on as well, typically the movement of a block of data to or from another memory in the system.

In fact, in these coarser-grain scheduling techniques there are only two resources: processors and memory. Opportunities for software pipelining arise in loops and recursive procedures. For example, consider a function F that is applied to a sequence of blocks of data, computing on one such block per iteration of a loop:

for $i = 1, n$ **do**
  $b \leftarrow load(B[i])$
  $F(b)$
  store(B[i],b)
endfor

To software pipeline this loop at a coarse granularity, while the function F is being applied to one block $B[i]$, the next block $B[i+1]$ can be loaded in parallel from its current location (e.g., in another processor's memory) and the previously computed block $B[i-1]$ can be stored (copied) to its final destination. In this way, three different iterations of the loop ($i-1$, i and $i+1$) are overlapped: the store from the previous iteration, the compute for the current iteration, and the load of the next iteration. It is also worth pointing out that coarse-grain software pipelining with two resources (compute and memory/communication) can also be thought of as automating in a compiler or dynamic scheduler the classical implementation techniques of double and triple buffering commonly used to improve performance in systems that must maintain high incoming or outgoing data rates, such as on-line transaction processing systems and network switches.

Brook, an early language for GPU programming, pioneered using these ideas for the static scheduling of streaming computations [BFH+04]. Subsequently, hierarchical software pipelining, where the static scheduling is done at multiple scales to match the memory hierarchy (e.g., between a CPU's RAM and the GPU's memory, and again between disk and the CPU's RAM) was developed for Sequoia [FHK+06]. Most recently, overlapped communication and computation has been realized in a dynamic scheduler more akin to hardware techniques for instruction scheduling in Legion [BTSA12].

To conclude, techniques for extracting instruction-level parallelism have not only had a major impact on the design of both hardware and low-level compilation, but have also begun to influence higher levels of the compilation stack. Given the growing diversity of hardware and the likelihood that this diversity will increase in the future as hardware vendors seek ever more efficient designs through specialization, as well as the widening gap that compilers must bridge between high-level programming languages and the underlying hardware, we expect that techniques for instruction level parallelism will find even more applications in the future.

Bibliography

[AAG+85] M. Annaratone, E. Arnould, T. Gross, H. T. Kung, and M. S. Lam. Warp architecture and implementation. In *Proceedings of the 12th International Symposium on Computer Architecture*, pages 346–356, 1985.

[AAG+87] M. Annaratone, E. Arnould, T. Gross, H. T. Kung, M. Lam, and O. Menzilcioglu. The Warp computer: Architecture, implementation and performance. *IEEE Transactions on Computers*, C-36(12):1523–1538, December 1987.

[ABJ64] G.M. Amdahl, G.A. Blaauw, and F.P. Brooks Jr. Architecture of the IBM System/360. *IBM Journal of Research and Development*, 8(2):87–101, April 1964.

[AC87] T. Agerwala and J. Cocke. High performance reduced instruction-set computers. Research Report RC-12434, IBM Thomas J. Watson Research Center, Yorktown Heights, NY, 1987.

[ACD74] T. L. Adam, K. M. Chandy, and J. R. Dickson. A comparison of list schedules for parallel processing systems. *Communications of the ACM*, 17(12):685–690, 1974.

[ACS+02] A. Aletá, J. M. Codina, F. J. Sánchez, A. González, and D. R. Kaeli. Exploiting pseudo-schedules to guide data dependence graph partitioning. In *Proceedings of the 2002 International Conference on Parallel Architectures and Compilation Techniques*, pages 281–290, 2002.

A. Aiken et al., *Instruction Level Parallelism*,
DOI 10.1007/978-1-4899-7797-7

[ACSG01] A. Aletá, J. M. Codina, J. Sánchez, and A. González. Graph-partitioning based instruction scheduling for clustered processors. In *Proceedings of the 34th Annual ACM/IEEE International Symposium on Microarchitecture*, pages 150–159, Austin, TX, 2001.

[Ada68] D. A. Adams. A computation model with data flow sequencing. Technical Report CS 119, Dept. of Computer Science, Stanford University, December 1968.

[Age76] T. Agerwala. Microprogram optimization: A survey. *IEEE Transactions on Computers*, C-25:962–973, October 1976.

[AGR95] E. R. Altman, R. Govindrajan, and G. R. Rao. Scheduling and mapping: software pipelining in the presence of hazards. In *Proceedings of the SIGPLAN '95 Conference on Programming Language Design and Implementation*, 1995.

[Aik88] A. S. Aiken. *Compaction-based parallelization*. PhD thesis, Dept. of Computer Science, Cornell University, August 1988.

[Aik90] A. Aiken. A theory of compaction-based parallelization. *Theoretical Computer Science*, 73(2):121–154, 1990.

[AJLA95] V. H. Allan, Reese B. Jones, Randall M. Lee, and Stephen J. Allan. Software pipelining. *ACM Computing Surveys*, 27(3):367–432, 1995.

[AK81] Arvind and V. Kathail. A multiple processor data flow machine that supports generalized procedures. In *Proceedings of the 8th Annual Symposium on Computer Architecture*, pages 291–302, Minneapolis, MN, 1981.

[AK01] R. Allen and K. Kennedy. *Optimizing Compilers for Modern Architectures: A Dependence-based Approach*. Morgan Kaufmann Publishers, 2001.

[AKPW83] J. R. Allen, K. Kennedy, C. Porterfield, and J. Warren. Conversion of control dependence to data dependence. In *Conference Record of the Tenth Annual ACM Symposium on the Principles of Programming Languages*, Austin, TX, January 1983.

[ALSU06] A. V. Aho, M. Lam, R. Sethi, and J. Ullman. *Compilers: Principles, Techniques, and Tools.* Addison-Wesley, Reading, MA, third edition, 2006.

[AM88] V. H. Allan and R. A. Mueller. Compaction with general synchronous timing. *IEEE Transactions on Software Engineering*, 14(5):595–599, 1988.

[Amd64] G. M. Amdahl. The structure of system/360, Part III: Processing unit design considerations. *IBM Journal of Research and Development*, 3(2):144–164, 1964.

[AN87] A. Aiken and A. Nicolau. Perfect pipelining: A new loop parallelization technique. Technical Report 87–873, Dept. of Computer Science, Cornell University, 1987.

[AN88] A. Aiken and A. Nicolau. Optimal loop parallelization. In *Proceedings of the SIGPLAN '88 Conference on Programming Language Design and Implementation*, Atlanta, GA, June 1988. ACM DOI 10.1145/960116.54021.

[AN90a] A. Aiken and A. Nicolau. Fine-grain parallelization and the wavefront method. In *Proceedings of the Third Workshop on Languages and Compilers for Parallel Computing*, Irvine, CA, 1990.

[AN90b] A. Aiken and A. Nicolau. A realistic resource-constrained software pipelining algorithm. In *Advances in Languages and Compilers for Parallel Computing*, Irvine, CA, August 1990. The MIT Press.

[AN95] A. Aiken and A. Nicolau. Resource constrained software pipelining. *IEEE Transactions on Parallel and Distributed Systems*, 6(12):1248–1270, 1995.

[Ana] SHARC Processors. http://www.analog.com/en/processors-dsp/sharc/products/index.html.

[ANvEB02] M. Agarwal, S. K. Nandy, J. v. Eijndhoven, and S. Balakrishnan. Speculative trace scheduling in VLIW processors. In *Proceedings of the 20th International Conference on Computer Design*, Freiburg, Germany, September 2002.

[Apo88] Apollo Computer, Inc., Chelmsford, MA. *The Series 10000 personal supercomputer: Inside a New Architecture, Publication No. 002402-007 2–88*, 1988.

[AR76] A. K. Agerwala and T. G. Rauscher. *Foundations of Microprogramming Architecture, Software, and Applications*. Academic Press, New York, NY, 1976.

[AST67] D. Anderson, F. Sparacio, and R. Tomasulo. The IBM System/360 model 91: Machine philosophy and instruction-handling. *IBM Journal of Research and Development*, 11(1):8–24, 1967.

[Atk91] M. Atkins. Performance and the i860 microprocessor. *IEEE Micro*, 11(5):24–27, 72–78, 1991.

[AWZ64] R. W. Allard, K. A. Wolf, and R. A. Zemlin. Effects of the 6600 computer on language structures. *Communications of the ACM*, 7(2):112–119, 1964.

[Bae80] J. L. Baer. *Computer Systems Architecture*. Computer Press, 1980.

[Bak74] K. R. Baker. *Introduction to sequencing and scheduling*. John Wiley and Sons, New York, NY, 1974.

[Ban76] U. Banerjee. Data dependence in ordinary programs. Master's thesis, Department of Computer Science, University of Illinois at Urbana-Champaign, November 1976. Report No. 76–837.

[Ban97] U. Banerjee. *Dependence Analysis*. Kluwer Academic Publishers, Boston, MA, 1997.

[Ban11a] U. Banerjee. Basic block parallelization. In D. Padua, editor, *Encyclopedia of Parallel Computing*, pages 1450–1458. Springer, 2011.

[Ban11b] U. Banerjee. Mathematical foundation of trace scheduling. *ACM Trans. Program. Lang. Syst.*, 33(3):10:1–10:24, May 2011.

[Bar78] G. E. Barnes. Comments on the identification of maximal parallelism in straight-line microprograms. *IEEE Transactions on Computers*, C-27(3):286–287, March 1978.

[Bar84] H.P. Barendregt. *The Lambda Calculus: Its Syntax and Semantics, VOLUME 103 of Studies in Logic and the Foundations of Mathematics.* North-Holland, Amsterdam, The Netherlands, 1984.

[BBGC86] W. A. Barrett, R. M. Bates, D. A. Gustafson, and J. D. Couch. *Compiler construction: Theory and Practice (2nd ed.).* SRA School Group, 1986.

[BBK+68] G. H. Barnes, R. M. Brown, M. Kato, D. J. Kuck, D. L. Slotnick, and R. A. Stokes. The ILLIAC IV computer. *IEEE Transactions on Computers*, C-17(8):746–757, August 1968.

[BCF+91] R. Bahr, S. Ciavaglia, B. Flahive, M. Kline, P. Mageau, and D. Nickel. The DN10000TX: A new high-performance PRISM processor. In *Proceedings of the COMPCON*, 1991.

[BCK91] D. Bernstein, D. Cohen, and H. Krawczyk. Code duplication: an assist for global instruction scheduling. In *Proceedings of the 24th International Symposium of Microarchitecture MICRO-24*, pages 103–113, 1991.

[BCS96] S. J. Beaty, S. Colcord, and P. H. Sweany. Using genetic algorithms to fine-tune instruction-scheduling heuristics. In *Proceedings of the 2nd International Conference on Massively Parallel Computing Systems (MPCS'96)*, Ischia, Italy, May 1996.

[Bea91] S. J. Beaty. Genetic algorithms and instruction scheduling. In *Proceedings of the 24th International Symposium of Microarchitecture MICRO-24*, pages 206–211, 1991.

[Bea92] S. J. Beaty. Lookahead scheduling. In *Proceedings of the 25th Annual International Symposium on Microarchitecture*, pages 256–259, 1992.

[Bel58] R. Bellman. On a routing problem. *Quarterly of Applied Mathematics*, 16(1):87–90, 1958.

[BFH+04] I. Buck, T. Foley, D. Horn, J. Sugerman, K. Fatahalian, M. Houston, and P. Hanrahan. Brook for GPUs: Stream computing on graphics hardware. In *ACM Transactions on Graphics*, volume 23, pages 777–786, 2004.

[BGM90] H. B. Bakoglu, G. F. Grohoski, and R. K. Montoye. The IBM RISC System/6000 processor: hardware overview. *IBM Systems Journal*, 34(1):12–22, 1990.

[BJS80] J. Bruno, J. W. Jones, and K. So. Deterministic scheduling with pipelined processors. *IEEE Computer*, C-29(4): 308–316, April 1980.

[BK92] G. Blanck and S. Krueger. The SuperSPARC™ microprocessor. In *Proceedings of the COMPCON*, pages 136–141, February 1992.

[Blo59] E. Block. The engineering design of the STRETCH computer. In *Proceedings of EJCC*, pages 48–59, 1959.

[BLO00] C. Basoglu, W. Lee, and J. Setel O'Donnell. The MAP1000A VLIW mediaprocessor. *IEEE Micro*, 20(2):48–59, March 2000.

[BR91] D. Bernstein and M. Rodeh. Global instruction scheduling for superscalar machines. In *Proceedings of the SIGPLAN '91 Conference on Programming Language Design and Implementation*, pages 241–255, 1991.

[BTSA12] M. Bauer, S. Treichler, E. Slaughter, and A. Aiken. Legion: Expressing locality and independence with logical regions. In *Proceedings of the Conference on Supercomputing*, pages 1–11, 2012.

[BTT99] S. Bharitkar, K. Tsuchiya, and Y. Takefuji. Microcode optimization with neural networks. *IEEE Transactions on Neural Networks*, 10(3):698–703, May 1999.

[BWJ90] S. Beaty, D. Whitley, and G. Johnson. Motivation and framework for using genetic algorithms for microcode compaction. In *Proceedings of the 23rd Annual Workshop and Symposium on Microprogramming and microarchitecture*, pages 117–124, Orlando, FL, 1990.

[BYA93] G. R. Beck, D. W. L. Yen, and T. L. Anderson. The Cydra 5 minisupercomputer: Architecture and implementation. *The Journal of Supercomputing*, 7(1–2):143–180, 1993.

[Cad] Tensilica Introduces Xtensa LX2 and Xtensa 7 Configurable Processors. http://ip.cadence.com/news/191/330/Tensilica-Introduces-Xtensa-LX2-and-Xtensa-7-Configurable-Processors.htm.

[CCK88] D. Callahan, J. Cocke, and K. Kennedy. Estimating interlock and improving balance for pipelined machines. *Journal of Parallel and Distributed Computing*, 5(4):334–358, August 1988.

[CDF64] B. B. Clayton, E. K. Dorff, and R. E. Fagen. An operating system and programming systems for the 6600. In *American Federation of Information Processing Societies (AFIPS) Proceedings of the Fall Joint Computer Conference (FJCC), part 2, 26*, pages 41–57, 1964.

[CF87] R. Cytron and J. Ferrante. What's in a name? or the value of renaming for parallelism detection and storage aladdress. In *Proceedings of the 1987 International Conference on Parallel Processing*, St. Charles, IL, August 1987.

[CG72] E. Coffman and R. Graham. Optimal scheduling for two processor systems. *Acta Informatica*, 1(3):200–213, 1972.

[CH88] P. P. Chang and W. W. Hwu. Trace selection for compiling large C application programs to microcode. In *Proceedings of the 21st Workshop on Microprogramming*, pages 21–29, San Diego, CA, 1988.

[Cha81] A. Charlesworth. An approach to scientific array processing: The architectural design of the AP-120B/FPS-164 family. *IEEE Computer*, 14(9):18–27, September 1981.

[CHL+03] H. Chen, W. C. Hsu, J. Lu, P. C. Yew, and D. Y. Chen. Dynamic trace selection using performance monitoring hardware sampling. In *Proceedings of the International Symposium on Code Generation and Optimization*, pages 79–90, San Francisco, CA, 2003.

[CK75] S. C. Chen and D. J. Kuck. Time and parallel processor bounds for linear recurrence systems. *IEEE Transactions on Computers*, C-24(7):701–717, July 1975.

[CLG02] J. M. Codina, J. Llosa, and A. Gonzalez. A comparative study of modulo scheduling techniques. In *Proceedings of the 16th ACM International Conference on Supercomputing*, UNC, Universitat Politechnica de Catalunya, June 2002.

[CLR90] T. H. Cormen, C. E. Leiserson, and R. L. Rivest. *Introduction to Algorithms*. The MIT Press, Cambridge, MA, 1990.

[CMM67] R. W. Conway, W. L. Maxwell, and L. W. Miller. *Theory of scheduling*. Addison-Wesley, Reading, MA, 1967.

[CNO+88] R. P. Colwell, R. P. Nix, J. O'Donell, D. B. Papworth, and P. K. Rodman. A VLIW architecture for a trace scheduling compiler. *IEEE Transactions on Computers*, 37(8):967–979, August 1988.

[Cof76] E. Coffman, Jr. *Computer and job-shop scheduling theory*. John Wiley and Sons, New York, NY, 1976.

[Cor79] M. Cornish. The TI dataflow architectures: The power of concurrency for avionics. In *Proceedings of the 3rd Conference on Digital Avionics Systems*, pages 19–25, Fort Worth, TX, November 1979.

[CS95] R. Colwell and R. Steck. A 0.6μm BiCMOS microprocessor with dynamic execution. In *Proceedings ISSCC '95 - International Solid-State Circuits Conference*, pages 176–177, San Francisco, CA, 1995.

[CSS98] K. Cooper, P. Schielke, and D. Subramanian. An experimental evaluation of list scheduling. Technical Report 98–326, Rice University, September 1998.

[CT03] K. D. Cooper and L. Torczon. *Engineering a Compiler*. Morgan Kaufmann Publishers, 2003.

[CTW86] B. E. Carpenter, A. M. Turing, and M. Woodger. A. M. Turing's ACE Report of 1946 and other papers. The MIT Press, Cambridge, MA, 1986.

[Cyt84] R. Cytron. *Compile-time Scheduling and Optimization for Asynchronous Machines.* PhD thesis, Department of Computer Science, University of Illinois at Urbana-Champaign, October 1984.

[DA92] K. Diefendorff and M. Allen. Organization of the Motorola 88110 superscalar RISC microprocessor. *IEEE Micro*, 12(2):40–63, March 1992.

[Das77] S. Dasgupta. Parallelism in loop-free microprograms. In *Information Processing '77, North-Holland*, pages 745–750, 1977.

[Dav69] R. L. Davis. The ILLIAC IV processing element. *IEEE Transactions on Computers*, C-18(9):800–816, September 1969.

[Dav71] E. S. Davidson. The design and control of pipelined function generators. In *Proceedings of International Conference on System Networks and Computers*, pages 19–21, 1971.

[Dav74] E. S. Davidson. Scheduling for pipelined processors. In *Proceedings of 7th Hawaii Conference on System Sciences*, pages 58–60, 1974.

[Dav78] A. L. Davis. The architecture and system method of DDM1: A recursively structured data driven machine. In *Proceedings of the 5th Annual Symposium on Computer architecture*, pages 210–215, 1978.

[DCS97] C. Ding, S. Carr, and P. H. Sweany. Modulo scheduling with cache reuse information. In *Proceedings of the Third International Euro-Par Conference on Parallel Processing*, pages 1079–1083, 1997.

[dD94] B. Dupont de Dinechin. Simplex scheduling: More than lifetime-sensitive instruction scheduling. In *Proceedings of the International Conference on Parallel Architectures and Compilation Techniques (PACT)*, Montreal, Canada, August 1994.

[Den74] J. B. Dennis. First version of a data-flow procedural language. In *Proceedings of the Colloque sur la Programmation, VOLUME 19 of Lecture Notes in Computer Science*, pages 362–376. Springer-Verlag, 1974.

[Den91] J. B. Dennis. The evolution of 'static' data-flow architecture. In J. L. Gaudiot and L. Bic, editors, *Advanced Topics in Data-Flow Computing, Chapter* 2. Prentice-Hall, 1991.

[DFL74] J. B. Dennis, J. B. Fosseen, and J. P. Linderman. Data flow schemas. In *Proceedings of the International Symposium on Theoretical Programming*, pages 187–216, 1974.

[DG83] J. B. Dennis and G. R. Gao. Maximum pipelining of array operations on static data flow machine. In *Proceedings of the 1983 International Conference on Parallel Processing*, pages 331–334, August 1983.

[DH96] D. A. Dunn and W. C. Hsu. Instruction scheduling for the HP PA-8000. In *Proceedings of the 29th International Symposium of Microarchitecture MICRO-29*, pages 298–307, Paris, France, 1996.

[DHB89] J. C. Dehnert, P. Y. T. Hsu, and J. P. Bratt. Overlapped loop support in the Cydra 5. In *Proceedings of the Third International Conference on Architectural Support for Programming Languages and Operating Systems (ASPLOS-III)*, pages 26–38, Boston, MA, 1989.

[DK82] A. L. Davis and R. M. Keller. Dataflow program graphs. *IEEE Computer*, 15(2):26–41, February 1982.

[DM75] J. B. Dennis and D. P. Misunas. A preliminary architecture for a basic data-flow processor. In *Proceedings of the 2nd Annual Symposium on Computer architecture*, pages 126–132, 1975.

[Don86] J. J. Dongarra. A survey of high performance computers. In *Proceedings of the COMPCON*, pages 8–11, March 1986.

[DSTP75] E. S. Davidson, L. E. Shar, A. T. Thomas, and J. H. Patel. Effective control for pipelined computers. In *Proceedings of COMPCON*, pages 181–184, San Francisco, CA, February 1975.

[DT76] S. Dasgupta and John Tartar. The identification of maximal parallelism in straight-line microprograms. *IEEE Transactions on Computers*, C-25(10):986–992, October 1976.

[DWYF92] E. DeLano, W. Walker, J. Yetter, and M. Forsyth. A high speed superscalar PA-RISC processor. In *Proceedings of the COMPCON*, pages 116–121, February 1992.

[Ebc87] K. Ebcioğlu. A compilation technique for software pipelining of loops with conditional jumps. In *Proceedings of the 20th Workshop on Microprogramming*, Colarado Springs, CO, December 1987.

[Ebc88] K. Ebcioğlu. Some design ideas for a VLIW architecture for sequential-natured software. In *Proceedings of IFIP WG 10.3 Working Conference on Parallel Processing*, pages 3–21, North-Holland, April 1988.

[ECTS59] J. P. Eckert, J. C. Chu, A. B. Tonik, and W. F. Schmitt. Design of UNIVAC-LARC system I. In *Proceedings of 1959 Eastern Joint Computer Conference*, pages 59–65, New York, NY, 1959.

[ED95] A. E. Eichenberger and Edward S. Davidson. Stage scheduling: A technique to reduce the register requirements of a modulo schedule. pages 338–349, 1995.

[Elba] Elbrus 8C with eight cores should be 250 GFlops reach. http://translate.yandex.com/translate?lang=de-en&url=http://www.golem.de/news/russischer-prozessor-elbrus-8c-mit-acht-kernen-soll-250-gflops-erreichen-1407-107869.html.

[Elbb] The Elbrus-2: a Soviet-era high performance computer. http://www.computerhistory.org/atchm/the-elbrus-2-a-soviet-era-high-performance-computer/.

[Ell85] J. R. Ellis. *Bulldog: A compiler for VLIW architectures (parallel computing, reduced-instruction-set, trace scheduling, scientific)*. PhD thesis, Dept. of Computer Science, Yale University, February 1985.

[EN89] K. Ebcioğlu and A. Nicolau. A global resource-constrained parallelization technique. In *Proceedings of the Third International Conference on Supercomputing*, pages 154–163, 1989.

[EN90] K. Ebcioğlu and T. Nakatani. A new compilation technique for parallelizing loops with unpredictable branches on a VLIW architecture. In *Proceedings of the Third Workshop on Languages and Compilers for Parallel Computing*, Urbana, IL, May 1990.

[FBF+00] Paolo Faraboschi, Geoffrey Brown, Joseph A. Fisher, Giuseppe Desoli, and Fred Homewood. Lx: A Technology Platform for Customizable VLIW Embedded Processing. In *Proceedings of the 27th Annual International Symposium on Computer Architecture*, pages 203–213, Vancouver, British Columbia, Canada, 2000.

[Fea88] P. Feautrier. Array expansion. In *Proceedings of the 2nd International Conference on Supercomputing*, St. Malo, France, July 1988.

[Fea91] P. Feautrier. Dataflow analysis of scalar and array references. *International Journal of Parallel Programming*, 20(1):23–52, February 1991.

[Fea94] P. Feautrier. Fine-grain scheduling under resource constraints. In *Proceedings of the Seventh Annual Workshop on Languages, Compilers and Compilers for Parallel Computers*, Ithaca, NY, August 1994.

[FERN84] J. A. Fisher, J. R. Ellis, J. C. Ruttenberg, and A. Nicolau. Parallel processing: A smart compiler and a dumb machine. *ACM SIGPLAN Notices*, 19(6):37–47, June 1984.

[FF92] J. A. Fisher and S. M. Freudenberger. Predicting conditional branch directions from previous runs of a program. In *Proceedings of the Fifth International Conference on Architectural Support for Programming Languages and Operating Systems (ASPLOS-V)*, pages 85–95, Boston, MA, 1992.

[FFY01] P. Faraboschi, J. A. Fisher, and C. Young. Instruction scheduling for instruction level parallel processors. *Proceedings of the IEEE*, 89(11):1638–1659, 2001.

[FFY04] J. A. Fisher, P. Faraboschi, and C. Young. *Embedded Computing: A VLIW Approach to Architecture, Compilers and Tools*. Morgan Kaufmann Publishers, 2004.

[FGL94] S. M. Freudenberger, T. R. Gross, and P. G. Lowney. Avoidance and suppression of compensation code in a trace scheduling compiler. *ACM Transactions on Programming Languages and Systems*, 16(4):1156–1214, July 1994.

[FHK+06] K. Fatahalian, D. Horn, T. Knight, L. Leem, M. Houston, J. Y. Park, M. Erez, M. Ren, A. Aiken, W. Dally, and P. Hanrahan. Sequoia: Programming the memory hierarchy. In *Proceedings of the Conference on Supercomputing*, 2006.

[Fis79] J. A. Fisher. *The Optimization of Horizontal Microcode Within and Beyond Basic Blocks: An Application of Processor Scheduling Beyond Basic Blocks*. PhD thesis, New York University, 1979.

[Fis81a] J. A. Fisher. Microcode compaction: Looking backward and looking forward. In *Proceedings of the National Computer Conference*, pages 95–102, Chicago, IL, July 1981.

[Fis81b] J. A. Fisher. Trace Scheduling: A technique for global microcode compaction. *IEEE Transactions on Computers*, C-30(7):478–490, July 1981.

[Fis84] J. Fisher. The VLIW machine: A multiprocessor for compiling scientific code. *IEEE Transactions on Computers*, 17(7):45–53, July 1984.

[Fis87] J. A. Fisher. VLIW architectures: Supercomputing via overlapped execution. In *Proceedings of the Second International Conference on Supercomputing*, Santa Barbara, CA, May 1987.

[Fis90] J. A. Fisher. VLIW architectures: an inevitable standard for the future? *Supercomputer*, 7(2):29–36, 1990.

[Fis93] J. A. Fisher. Global code generation for instruction-level parallelism: Trace Scheduling-2. Technical Report HPL-93-43, Hewlett Packard Laboratories, June 1993.

[FLT99] M. M. Fernandes, J. Llosa, and N. Topham. Distributed modulo scheduling. In *Proceedings of the Fifth International Symposium on High-Performance Computer Architecture*, page 130, 1999.

[FO84] J. A. Fisher and J. J. O'Donnell. VLIW Machines: Multiprocessors we can actually program. In *Spring CompCon 84 Digest of Papers*, February 1984.

[FR72] C. C. Foster and E. M. Riseman. Percolation of code to enhance parallel dispatching and execution. *IEEE Transactions on Computers*, C-21(12):1411–1415, December 1972.

[FSP79] Floating Point Systems, Inc., Beaverton, OR. *FPS AP-120B Processor HandBOOK*, 1979.

[Ful98] S. Fuller. *Motorola's AltiVec™ Technology*. Freescale Semiconductor, Inc, January 1998.

[GAG94] R. Govindarajan, Erik R. Altman, and Guang R. Gao. Minimizing register requirements under resource-constrained rate-optimal software pipelining. In *Proceedings of the 27th International Symposium of Microarchitecture*, pages 85–94, San Jose, CA, 1994.

[GAG96] R. Govindarajan, Erik R. Altman, and Guang R. Gao. A framework for resource-constrained rate-optimal software pipelining. *IEEE Transactions on Parallel and Distributed Systems*, 7(11):1133–1149, 1996.

[GBC+95] D. Greenley, J. Bauman, D. Chang, D. Chen, R. Eltejaein, P. Ferolito, P. Fu, R. Garner, D. Greenhill, H. Grewal, K. Holdbrook, B. Kim, L. Kohn, H. Kwan, M. Levitt, G. Maturana, D. Mrazek, C. Narasimhaiah, K. Normoyle, N. Parveen, P. Patel, A. Prabhu, M. Tremblay, M. Wong, L. Yang, K. Yarlagadda, R. Yu, R. Yung, and G. Zyner. UltraSPARC™ : The next generation superscalar 64-bit SPARC. In *In Proceedings of COMPCON '95*, pages 442–451, March 1995.

[GDHH89] V. G. Grafe, G. S. Davidson, J. E. Hoch, and V. P. Holmes. The Epsilon dataflow processor. In *Proceedings of the 16th Annual International Symposium on Computer architecture*, pages 36–45, Jerusalem, Israel, 1989.

[Gep02] Pawel Gepner. Overview of ia-64 explicitly parallel instruction computing architecture. In Roman Wyrzykowski, Jack Dongarra, Marcin Paprzycki, and Jerzy Waniewski, editors, *Parallel Processing and Applied Mathematics*, volume 2328 of *Lecture Notes in Computer Science*, pages 331–339. 2002.

[GFO92] A. De Gloria, P. Faraboschi, and M. Olivieri. A nondeterministic scheduler for a software pipelining compiler. In *Proceedings of the 25th International Symposium of Microarchitecture*, pages 41–44, 1992.

[Gir91] M. B. Girkar. *Functional Parallelism Theoretical Foundations and Implementation.* PhD thesis, Department of Computer Science, University of Illinois at Urbana-Champaign, December 1991.

[GJ79] M. Garey and D. Johnson. *Computers and Intractability, A Guide to the Theory of NP-Completeness.* W. H. Freeman and Co., New York, NY, 1979.

[GKLW85] T. Gross, H. T. Kung, M. Lam, and J. Webb. Warp as a machine for low-level vision. In *Proceedings of IEEE International Conference on Robotics and Automation*, pages 790–800, March 1985.

[GKW85] J. R Gurd, C. C Kirkham, and I. Watson. The Manchester prototype dataflow computer. *Communications of the ACM*, 28(1):34–52, January 1985.

[GM63] J. Gregory and R. McReynolds. The SOLOMON computer. *IEEE Transactions on Electronic Computers*, EC-12:774–781, December 1963.

[GM86] P. B. Gibbons and S. S. Muchnick. Efficient instruction scheduling for a pipelined architecture. In *Proceedings of the 1986 SIGPLAN Symposium on Compiler contruction*, pages 11–16, Palo Alto, CA, 1986.

[GO98] T. Gross and D. R. O'Hallaron. *iWarp: Anatomy of a Parallel Computing System*. The MIT Press, Cambridge, MA, March 1998.

[Gon77] M. J. Gonzalez. Deterministic processor scheduling. *ACM Computing Surveys*, 9(3):173–204, September 1977.

[GP92] M. Girkar and C. D. Polychronopoulos. Automatic extraction of functional parallelism from ordinary programs. *IEEE Transactions on Parallel and Distributed Systems*, 3(2):166–178, 1992.

[Gro83] T. R. Gross. Code optimization techniques for pipelined architectures. In *Proceedings of the 1983 Spring COMPCON*, pages 278–285, San Francisco, CA, March 1983.

[Gro90] G.F. Grohoski. Machine organization of the IBM RISC System/6000 processor. *IBM Systems Journal*, 34(1): 37–58, 1990.

[GS83] R. Grishman and B. Su. A preliminary evaluation of trace scheduling for global microcode compaction. *IEEE Transactions on Computers*, C-32(12):1191–1194, December 1983.

[GS94] F. Gasperoni and U. Schwiegelshohn. Generating close to optimum loop schedules on parallel processors. *Parallel Processing Letters*, 4(4):391–403, 1994.

[GSE89] F. Gasperoni, U. Schwiegelshohn, and K. Ebcioğlu. On optimal loop parallelization. In *Proceedings of the 22nd Workshop on Microprogramming*, pages 141–147, 1989.

[GW90] T. R. Gross and M. Ward. The suppression of compensation code. In *Proceedings of the Third Workshop on Languages and Compilers for Parallel Computing*, pages 260–273, Palo Alto, CA, 1990.

[Har87] M. Harris. Extending microcode compaction for real architectures. In *Proceedings of the 20th Annual Workshop on Microprogramming*, pages 40–53, Colorado Springs, CO, 1987.

[HBK+03] K. Heydemann, F. Boldin, P. M. W. Knijnenburg, L. Morin, and H. Charles. UFC: A global trade-off strategy for loop unrolling for VLIW architecture. In *Proceedings of the Tenth Workshop on Compilers for Parallel Computers*, Amsterdam, The Netherlands, January 2003.

[HG82] J. L. Hennessy and T. R. Gross. Code generation and reorganization in the presence of pipeline constraints. In *Proceedings of the 9th ACM SIGPLAN-SIGACT Symposium on Principles of programming languages*, pages 120–127, Albuquerque, Mexico, 1982.

[Hig78] L. C. Higbie. Overlapped operation with microprogramming. *IEEE Transactions on Computers*, 27(3):270–275, 1978.

[HJ76] J. L. Hilburn and P. N. Julich. *Microcomputers/Microprocessors: Hardware, Software, and Applications*. Prentice-Hall, Englewood Cliff, NJ, 1976.

[HJ81] R. W. Hockney and C. R. Jesshope. *Parallel Computers: Architecture, Programming and Algorithms*. IOP Publishing Ltd., Bristol, UK, 1981.

[HJ88] R. W. Hockney and C. R. Jesshope. *Parallel Computers 2: Architecture, Programming and Algorithms*. IOP Publishing Ltd., Bristol, UK, 1988.

[HJP+82] J. Hennessy, N. Jouppi, S. Przybylski, C. Rowen, T. Gross, F. Baskett, and J. Gill. MIPS: A microprocessor architecture. In *Proceedings of the 15th Annual Workshop on Microprogramming*, pages 17–22, Palo Alto, CA, 1982.

[HMC+93] W. M. W. Hwu, S. A. Mahlke, W. Y. Chen, P. P. Chang, N. J. Warter, R. A. Bringmann, R. G. Ouellette, R. E. Hank, T. Kiyohara, G. E. Haab, J. G. Holm, and D. M. Lavery. The superblock: An effective technique for VLIW and superscalar compilation. *The Journal of Supercomputing*, 7(1–2):229–248, November 1993.

[HNI82] E. Hogenauer, R. F. Newbold, and Y. J. Inn. DDSP – a data flow computer for signal processing. In *Proceedings of the 1982 International Conference on Parallel Processing*, pages 126–133, 1982.

[HP90] J. Hennessy and D. Patterson. *Computer Architecture A Quantitative Approach*. Morgan Kaufmann Publishers, San Mateo, CA, 1990.

[Hsu86] P. S. Hsu. *Highly Concurrent Scalar Processing*. PhD thesis, Department of Computer Science, University of Illinois at Urbana-Champaign, 1986.

[HT72] R. G. Hintz and D. P. Tate. Control Data STAR-100 processor design. In *Proceedings of the COMPCON*, pages 221–228, September 1972.

[HT73] J. Hopcroft and R. Tarjan. Algorithm 447: efficient algorithms for graph manipulation. *Communications of the ACM*, 16(6):372–378, 1973.

[Hu61] T. C. Hu. Parallel sequencing and assembly line problems. *Operations Research*, 9(6):841–848, 1961.

[Huf93] R. A. Huff. Lifetime-sensitive modulo scheduling. In *Proceedings of the SIGPLAN '93 Conference on Programming Language Design and Implementation*, pages 258–267, Albuquerque, NM, June 1993.

[Hus70] S. S. Husson. *Microprogramming : Principles and Practices*. Prentice-Hall, Englewood Cliff, NJ, 1970.

[Hwa93] K. Hwang. *Advanced Computer Architecture: Parallelism,Scalability,Programmability*. McGraw-Hill, 1993.

[IBM76] IBM Corporation, Endicott, NY. *IBM 3838 Array Processor Functional Characteristics, Publication No. 6A24-3639-0, File No. S370-08*, 1976.

[IKI83] S. Isoda, Y. Kobayaski, and T. Ishida. Global compaction of horizontal microprograms based on generalized data dependency graph. *IEEE Transactions on Computers*, C-32(10):922–933, October 1983.

[Inf] Infineon introduces configurable CARMEL DSP Core for 3G wireless and broadband communication applications. http://www.infineon.com/cms/en/corporate/press/news/releases/2000/129311.html.

[INT89] Intel Corporation, Santa Clara, CA. *i860 64-bit Microprocessor Programmer's Reference Manual, Publication No. 240329-001*, 1989.

[JA90] R. B. Jones and V. H. Allan. Software pipelining : A comparison and improvement. In *Proceedings of the 23rd Workshop on Microprogramming and Microarchitecture*, Orlando, FL, November 1990.

[Jai91] S. Jain. Circular scheduling: a new technique to perform software pipelining. In *Proceedings of the SIGPLAN '91 Conference on Programming Language Design and Implementation*, pages 219–228, Toronto, Canada, 1991.

[JD74] L. W. Jackson and S. Dasgupta. The identification of parallel microoperations. *Information Processing Letters*, 2:180–184, March 1974.

[Joh91] M. Johnson. *Superscalar Microprocessor Design*. Prentice-Hall, New Jersey, 1991.

[Jou89] N. P. Jouppi. The non-uniform distribution of instruction-level and machine parallelism and its effect on performance. *IEEE Computer*, C-38(12):1645–1658, 1989.

[JPSW82] D. Jacobs, J. Prins, P. Siegel, and K. Wilson. Monte carlo techniques in code optimization. In *Proceedings of the 15th Workshop on Microprogramming*, pages 143–148, 1982.

[JRS97] Q. Jacobson, E. Rotenberg, and J. E. Smith. Path-based next trace prediction. In *Proceedings of the 30th International Symposium of Microarchitecture*, pages 14–23, Research Triangle Park, NC, 1997.

[JW89] N. Jouppi and D. Wall. Available instruction-level parallelism for superscalar and superpipelined machines. In *Proceedings of the Third International Conference on Architectural Support for Programming Languages and Operating Systems (ASPLOS-III)*, pages 272–282, Boston, MA, April 1989.

[Kat85] M. G. H. Katevenis. *Reduced Instruction Set Computer Architectures for VLSI*. The MIT Press, 1985 1985.

[Kel75] R. M. Keller. Look-ahead processors. *ACM Computing Surveys*, 7(4):177–195, December 1975.

[KKP+81] D. Kuck, R. Kuhn, D. Padua, B. Leasure, and M. J. Wolfe. Dependence graphs and compiler optimizations. In *Conference Record of the Eighth Annual ACM Symposium on the Principles of Programming Languages*, Williamsburg, VA, January 1981.

[KM66] R. Karp and R. Miller. Properties of a model for parallel computation: determinacy, termination, queueing. *SIAM Journal of Applied Mathematics*, 4(6):1390–1411, November 1966.

[KM89a] L. Kohn and N. Margulis. Introducing the Intel i860 64-bit microprocessor. *IEEE Micro*, 9(4):15–30, July 1989.

[KM89b] Les Kohn and Neal Margulis. Introducing the Intel I860 64-Bit Microprocessor. *IEEE Micro*, 9(4):15–30, July 1989.

[KN85] K. Karplus and A. Nicolau. Efficient hardware for multiway jumps and pre-fetches. In *Proceedings of the 18th Annual Workshop on Microprogramming*, pages 11–18, 1985.

[KN92] K. Kim and A. Nicolau. N-dimensional perfect pipelining. Technical report, Computer Science Department, University of California at Irvine, 1992.

[KN11] A. Kejariwal and A. Nicolau. Modulo scheduling and loop pipelining. In D. Padua, editor, *Encyclopedia of Parallel Computing*, pages 1158–1173. Springer, 2011.

[Kog77a] P. Kogge. Algorithm development for pipelined processors. In *Proceedings of the 1977 International Conference on Parallel Processing*, August 1977.

[Kog77b] P. Kogge. The microprogramming of pipelined processors. In *Proceedings of the Fourth International Symposium on Computer Architecture*, pages 63–69, March 1977.

[Kog81] P. M. Kogge. *The Architecture of Pipelined Computers.* Hemisphere Publishing Corporation, 1981.

[Kou73] J. Koudela. The past, present, and future of minicomputers: A scenario. *Proceedings of the IEEE*, 61(11):1526–1534, November 1973.

[KSPS83] M. G. H. Katevenis, R. W. Sherburne, D. A. Patterson, and C. H. Sèquin. The RISC II micro-architecture. In F. Anceau and E. J. Aas, editors, *Proceedings of the VLSI 83 Conference*, pages 349–359. Elsevier Science Publishers (IFIP), North-Holland, 1983.

[KT79] J. Kim and C. J. Tan. Register assignment algorithms for optimizing micro-code compliers - Part I. Technical Report RC 4633 (# 20545), IBM T. J. Watson Research Center, May 1979.

[Kuh80] R. Kuhn. *Optimization and Interconnection Complexity for: Parallel Processors, Single-Stage Networks, and Decision Trees.* PhD thesis, Department of Computer Science, University of Illinois at Urbana-Champaign, February 1980.

[Kum97] A. Kumar. The HP PA-8000 RISC CPU. *IEEE Micro*, 17(2):27–32, 1997.

[Lam74] L. Lamport. The parallel execution of DO loops. *Communications of the ACM*, 17(2):83–93, February 1974.

[Lam88] M. Lam. Software pipelining: An effective scheduling technique for VLIW machines. In *Proceedings of the SIGPLAN '88 Conference on Programming Language Design and Implementation*, Atlanta, GA, June 1988.

[Lar73] J. L. Larson. Cost-effective processor design with an application to FFT. Technical Report SU-SEL-73-037, Dept. of Computer Science, Stanford University, August 1973.

[LAV98] J. Llosa, E. Ayguadé, and M. Valero. Quantitative evaluation of register pressure on software pipelined loops. *International Journal of Parallel Programming*, 26(2):121–142, 1998.

[Law76] E. L. Lawler. *Combinatorial Optimization: Networks and Matroids*. Holt, Rinehart and Winston, 1976.

[LB80] S. F. Lundstrom and G. H. Barnes. A controllable MIMD architectures. In *Proceedings of the 1980 International Conference on Parallel Processing*, pages 19–27, St. Charles, IL, August 1980.

[LDSM80] D. Landskov, S. Davidson, B.D. Shriver, and P.W. Mallett. Local microcode compaction techniques. *ACM Computing Surveys*, 12(3):261294, September 1980.

[LF02] J. Llosa and S. M. Freudenberger. Reduced code size modulo scheduling in the absence of hardware support. In *Proceedings of the 35th Annual ACM/IEEE International Symposium on Microarchitecture*, pages 99–110, Istanbul, Turkey, 2002.

[LFK+93] P. G. Lowney, S. M. Freudenberger, T. J. Karzes, W. D. Lichtenstein, R. P. Nix, J. S. O'Donnell, and J. Ruttenberg. The Multiflow trace scheduling compiler. *The Journal of Supercomputing*, 7(1–2):51–142, May 1993.

[LH95] D. M. Lavery and W. W. Hwu. Unrolling-based optimizations for modulo scheduling. In *Proceedings of the 28th Annual International Symposium on Microarchitecture*, pages 327–337, Ann Arbor, MI, 1995.

[LH96] D. M. Lavery and W. W. Hwu. Modulo scheduling of loops in control-intensive non-numeric programs. In *Proceedings of the 29th International Symposium of Microarchitecture MICRO-29*, pages 126–137, Paris, France, 1996.

[Lin83] J. L. Linn. SRDAG compaction: A generalization of trace scheduling to increase the use of global context information. In *Proceedings of the 16th Workshop on Microprogramming*, October 1983.

[Lin88] J. L. Linn. Horizontal microcode compaction. In S. Habib, editor, *Microprogramming and Firmware Engineering Methods*, pages 381–431. Van Nostrand Reinhold, New York, NY, 1988.

[LS90] J. Labrousse and G. Slavenburg. A 50 MHz microprocessor with a VLIW architecture. In *Proceedings of the International Solid State Circuits Conference*, San Francisco, CA, 1990.

[Luc91] E. Lucas. *Récréations Mathématiques*. Gauthier-Villares, Paris, France, 1891.

[Lun75] A. Lunde. *Evaluation of instruction set processor architecture by program tracing*. PhD thesis, 1975.

[Lun77] A. Lunde. Empirical evaluation of some features of instruction set processor architectures. *Communications of the ACM*, 20(3):143–153, March 1977.

[MCB+93] S. A. Mahlke, W. Y. Chen, R. A. Bringmann, R. E. Hank, W. W. Hwu, B. R. Rau, and M. S. Schlansker. Sentinel scheduling: a model for compiler-controlled speculative execution. *ACM Transactions on Computer Systems*, 11(4):376–408, 1993.

[McG90] S. McGeady. The i960CA superscalar implementation of the 80960 architecture. In *Proceedings of the 35th IEEE COMPCON*, pages 232–239, March 1990.

[MCGH92] S. A. Mahlke, W. Y. Chen, J. C. Gyllenhaal, and W.-M. W. Hwu. Compiler code transformations for superscalar-based high performance systems. In *Proceedings of the 1992 ACM/IEEE Conference on Supercomputing*, pages 808–817, 1992.

[MCH+92] S. A. Mahlke, W. Y. Chen, W. W. Hwu, B. R. Rau, and M. S. Schlansker. Sentinel scheduling for VLIW and superscalar processors. *ACM SIGPLAN Notices*, 27(9):238–247, 1992.

[MD76] P. Mateti and N. Deo. On algorithms for enumerating all circuits of a graph. *SIAM Journal of Computing*, 5(1):90–99, 1976.

[MDO84] R. A. Mueller, M. R. Duda, and S. M. O'Haire. A survey of resource allocation methods in optimizing microcode compilers. In *Proceedings of the 17th Annual Workshop on Microprogramming*, pages 285–295, 1984.

[ME92] S. M. Moon and K. Ebcioğlu. An efficient resource-constrained global scheduling technique for superscalar and VLIW processors. In *Proceedings of the 25th Annual International Symposium on Microarchitecture*, pages 55–71, 1992.

[MH01] M. C. Merten and W. W. Hwu. Modulo schedule buffers. In *Proceedings of the 34th Annual ACM/IEEE International Symposium on Microarchitecture*, pages 138–149, Austin, TX, 2001.

[MJ98] D. Milicev and Z. Jovanovic. Predicated software pipelining technique for loops with conditions. In *Proceedings of the 12th International Parallel Processing Symposium*, Orlando, FL, March 1998.

[MLC+92] S. A. Mahlke, D. C. Lin, W. Y. Chen, R. E. Hank, and R. A. Bringmann. Effective compiler support for predicated execution using the hyperblock. In *Proceedings of the 25th International Symposium of Microarchitecture*, pages 45–54, 1992.

[MPC05] MPC7450 RISC Microprocessor Family Reference Manual. http://www.datasheetarchive.com/dl/Datasheets-SW1/DSASW0013006.pdf, January 2005.

[MPF82] M. Mezzalama, P. Prinetto, and G. Filippi. Microcode compaction via microblock definition. In *Proceedings of the 15th Annual Workshop on Microprogramming*, pages 134–142, Palo Alto, CA, 1982.

[MPS87] T. E. Mankovich, V. Popescu, and H. Sullivan. Chopp principles of operation. In *Proceedings of the Second International Conference on Supercomputing*, pages 2–10, 1987.

[Muc00] S. Muchnick. *Advanced Compiler Design Implementation.* Morgan Kaufmann Publishers, second edition, 2000.

[Mur71] Y. Muraoka. *Parallelism Exposure and Exploitation in Programs.* PhD thesis, Department of Computer Science, University of Illinois at Urbana-Champaign, February 1971. Report No. 71–424.

[NE89] T. Nakatani and K. Ebcioğlu. "Combining" as a compilation technique for VLIW architectures. In *Proceedings of the 22nd Workshop on Microprogramming*, pages 43–55, 1989.

[NE98] E. Nystrom and A. E. Eichenberger. Effective cluster assignment for modulo scheduling. In *Proceedings of the 31st Annual ACM/IEEE International Symposium on Microarchitecture*, pages 103–114, Dallas, TX, 1998.

[NF81] A. Nicolau and J. A. Fisher. Using an oracle to measure potential parallelism in single instruction stream programs. In *Proceedings of the 14th Workshop on Microprogramming*, pages 171–182, Chatham, MA, 1981.

[NG93] Q. Ning and G. R. Gao. A novel framework of register allocation for software pipelining. In *Proceedings of the Twentieth Annual ACM Symposium on the Principles of Programming Languages*, pages 29–42, March 1993.

[Nic84] A. Nicolau. *Parallelism, memory anti-aliasing and correctness for trace scheduling compilers (disambiguation, flow-analysis, compaction).* PhD thesis, Dept. of Computer Science, Yale University, 1984.

[Nic85a] A. Nicolau. Loop quantization: Unwinding for fine-grain parallelism exploitation. Technical Report TR85-709, Dept. of Computer Science, Cornell University, October 1985.

[Nic85b] A. Nicolau. Percolation scheduling. In *Proceedings of the 1985 International Conference on Parallel Processing*, August 1985.

[Nic85c] A. Nicolau. Percolation scheduling : A parallel compilation technique. Technical Report TR85-678, Dept. of Computer Science, Cornell University, May 1985.

[Nic86] A. Nicolau. Percolation scheduling : A hierarchical parallel compilation technique. Technical report, Dept. of Computer Science, Cornell University, 1986.

[Nic87] A. Nicolau. Loop quantization or unwinding done right. In *Proceedings of the First International Conference on Supercomputing*, pages 294–308, Athens, Greece, June 1987. Springer-Verlag.

[Nic88] A. Nicolau. Loop quantization: a generalized loop unwinding technique. *Journal of Parallel and Distributed Computing*, 5(5):568–586, 1988.

[NN95] S. Novack and A. Nicolau. Trailblazing: A hierarchical approach to percolation scheduling. *International Journal of Parallel Programming*, 23(1), 1995.

[NN97] S. Novack and A. Nicolau. Resource directed loop pipelining. *IEEE Computer*, 40(6):311–321, 1997.

[NNN+91] M. Nakajima, H. Nakano, Y. Nakakura, T. Yoshida, Y. Goi, Y. Nakai, R. Segawa, T. Kishida, and H. Kadota. OHMEGA: A VLSI superscalar processor architecture for numerical applications. In *Proceedings of the 18th International Symposium on Computer Architecture*, pages 160–168, Toronto, Ontario, Canada, 1991.

[Pap96] D. B. Papworth. Tuning the Pentium Pro microarchitecture. *IEEE Micro*, 16(2):8–15, 1996.

[Pat78] J. H. Patel. Pipelines with internal buffers. In *Proceedings of the 5th Annual Symposium on Computer architecture*, pages 249–254, 1978.

[Pat85] David A. Patterson. Reduced instruction set computers. *Communications of the ACM*, 28(1):8–21, January 1985.

[PC90] G. M. Papadopoulos and D. E. Culler. Monsoon: an explicit token-store architecture. In *Proceedings of the 17th Annual Symposium on Computer Architecture*, pages 293–302, 1990.

[PD76] J. H. Patel and E. S. Davidson. Improving the throughput of a pipeline by insertion of delays. In *Proceedings of the Third International Symposium on Computer Architecture*, pages 159–164, 1976.

[Pet01] D. Petkov. Efficient pipelining of nested loops: Unroll and squash. Master's thesis, Massachusetts Institute of Technology, January 2001.

[PK89] P. Paulin and J. Knight. Force-directed scheduling for the behavioral synthesis of ASIC's. *IEEE Transactions on Computer-Aided Design of Integrated Circuits and Systems*, 8(6):661–679, 1989.

[PKL80] D. A. Padua, D. J. Kuck, and D. H. Lawrie. High-speed multiprocessors and compilation techniques. *IEEE Transactions on Computers*, 29(9):763–776, September 1980.

[Pla76] A. Plas. LAU system architecture: A parallel data driven processor based on single assignment. In *Proceedings of the 1976 International Conference on Parallel Processing*, pages 293–302, August 1976.

[Poe80] M. D. Poe. Heuristics for the global optimization of microprograms. In *Proceedings of the 13th Annual Workshop on Microprogramming*, pages 13–22, Springs, CO, 1980.

[PS81] D. A. Patterson and C. H. Sequin. RISC I: A Reduced Instruction Set VLSI computer. In *Proceedings of the 8th Annual Symposium on Computer Architecture*, pages 443–457, Minneapolis, MN, 1981.

[PS82] D. A. Patterson and C. H. S'equin. A VLSI RISC. *IEEE Computer*, 15(9):8–22, September 1982.

[PS91] J. Park and M. Schlansker. On predicated execution. Technical Report 58–91, Hewlett Packard Laboratories, 1991.

[PSS+91] V. Popescu, M. Schultz, J. Spracklen, G. Gibson, B. Lightner, and D. Isaman. The metaflow architecture. *IEEE Micro*, 11(3):10–13, 63–73, May 1991.

[RAB+12] R. Riedlinger, R. Arnold, L. Biro, B. Bowhill, J. Crop, K. Duda, E.S. Fetzer, O. Franza, T. Grutkowski, C. Little, C. Morganti, G. Moyer, A. Munch, M. Nagarajan, C. Parks, C. Poirier, B. Repasky, E. Roytman, T. Singh, and M.W. Stefaniw. A 32 nm, 3.1 Billion Transistor, 12 Wide Issue Itanium®Processor for Mission-Critical Servers. *Solid-State Circuits, IEEE Journal of*, 47(1):177–193, Jan 2012.

[Rad82] G. Radin. The 801 minicomputer. *ACM SIGPLAN Notices*, 17(4):39–47, April 1982.

[Ram94] J. Ramanujam. Optimal software pipelining of nested loops. In *Proceedings of the 8th International Symposium on Parallel Processing*, pages 335–342, 1994.

[Rau88] B. R. Rau. Cydra 5 directed dataflow architecture. In *Proceedings of COMPCON Spring'88*, pages 106–113, San Francisco, CA, 1988.

[Rau92] B. R. Rau. Data flow and dependence analysis for instruction level parallelism. In *Proceedings of the Fourth Workshop on Languages and Compilers for Parallel Computing*, pages 236–250, 1992.

[Rau95] B. R. Rau. Iterative modulo scheduling. Technical Report HPL-94-115, Hewlett Packard Laboratories, November 1995.

[RB69] J. E. Rodrigues and J. E. R. Bezos. A graph model for parallel computations. Technical Report 64, Massachusetts Institute of Technology, 1969.

[RBS96] E. Rotenberg, S. Bennett, and J. E. Smith. Trace cache: a low latency approach to high bandwidth instruction fetching. In *Proceedings of the 29th Annual ACM/IEEE International Symposium on Microarchitecture*, pages 24–35, Paris, France, 1996.

[RBS99] E. Rotenberg, S. Bennett, and J. E. Smith. A trace cache microarchitecture and evaluation. *IEEE Transactions on Computers*, 48(2):111–120, February 1999.

[RC69] J. F. Ruggiero and D. A. Coryell. An auxiliary processing system for array calculations. *IBM Systems Journal*, 8(2):118–135, 1969.

[RCG72] C. V. Ramamoorthy, K. M. Chandy, and M. J. Gonzalez. Optimal scheduling strategies in a multiprocessor system. *IEEE Transactions on Computers*, C-21(2):137–146, February 1972.

[Ree93] C. R. Reeves. *Modern Heuristic techniques for Combinatorial Problems*. John Wiley and Sons, New York, NY, 1993.

[RF72] E. M. Riseman and C. C. Foster. The inhibition of potential parallelism by conditional jumps. *IEEE Transactions on Computers*, C-21(12):1405–1411, December 1972.

[RF93] B. R. Rau and J. A. Fisher. Instruction level parallel processing: History, overview and perspective. *The Journal of Supercomputing*, 7(1):97, January 1993.

[RG81] B. R. Rau and C. D. Glaeser. Some scheduling techniques and an easily schedulable horizontal architecture for high performance scientific computing. In *Proceedings of the 14th Annual Workshop on Microprogramming*, pages 183–198, Chatham, MA, December 1981.

[RGP82] B. R. Rau, C. D. Glaeser, and R. L. Picard. Efficient code generation for horizontal architectures: Compiler techniques and architectural support. In *Proceedings of the Ninth International Symposium on Computer Architecture*, pages 131–139, Austin, TX, 1982.

[RGSL96] J. Ruttenberg, G. R. Gao, A. Stoutchinin, and W. Lichtenstein. Software pipelining showdown: Optimal vs. heuristic methods in a production compiler. In *Proceedings of the SIGPLAN '96 Conference on Programming Language Design and Implementation*, pages 1–11, Philadelphia, PA, 1996.

[RJSS97] E. Rotenberg, Q. Jacobson, Y. Sazeides, and J. Smith. Trace processors. In *Proceedings of the 30th Annual ACM/IEEE International Symposium on Microarchitecture*, pages 138–148, Research Triangle Park, NC, 1997.

[RL77] C. V. Ramamoorthy and H. F. Li. Pipeline architecture. *ACM Computing Surveys*, 9(1), 1977.

[RLTS92] B. R. Rau, M. Lee, P. P. Tirumalai, and M. S. Schlansker. Register allocation for software pipelined loops. *ACM SIGPLAN Notices*, 27(7):283–299, 1992.

[RS99] E. Rotenberg and J. Smith. Control independence in trace processors. In *Proceedings of the 32nd Annual ACM/IEEE International Symposium on Microarchitecture*, pages 4–15, Haifa, Israel, 1999.

[RST92] B. R. Rau, M. S. Schlansker, and P. P. Tirumalai. Code generation schema for modulo scheduled loops. In *Proceedings of the 25th International Symposium of Microarchitecture*, pages 158–169, Portland, OR, 1992.

[Rus78] R. M. Russell. The CRAY-1 computer system. *Communications of the ACM*, 21(1):63–72, January 1978.

[Rym82] J. W. Rymarczyk. Coding guidelines for pipelined processors. In *Proceedings of the First Symposium on Architectural Support for Programming Languages and Operation Systems (ASPLOS)*, pages 12–19, Palo Alto, CA, 1982.

[RYYT89] B. R. Rau, W. L. Yen, W. Yen, and A. Towle. The Cydra 5 departmental supercomputer: design philosophies, decisions and trade-offs. *IEEE Computer*, 22(1):12–35, January 1989.

[SA00] H. Sharangpani and K. Arora. Itanium processor microarchitecture. *IEEE Micro*, 20(5):24–43, September 2000.

[SB92] P. H. Sweany and S. J. Beaty. Dominator-path scheduling: A global scheduling method. In *Proceedings of the 25th International Symposium of Microarchitecture*, pages 260–263, 1992.

[SBN82] D. P. Siewiorek, C. G. Bell, and A. Newell. *Computer Structures: Principles and Examples*. McGraw-Hill, New York, NY, 1982.

[SC93] F. Sánchez and J. Cortadella. Resource-constrained pipelining based on loop transformations. *Microprocessing and Microprogramming*, 38(1–5):429–436, September 1993.

[SC94] F. Sánchez and J. Cortadella. UNRET: A transformation-based technique for software pipelining with resource constraints. Technical Report UPC-DAC-94-11, Departament d'Arquitectura de Computadors, Universitat Polytecnica de Catalunya, 1994.

[SC95] F. Sánchez and J. Cortadella. Time-constrained loop pipelining. Technical Report UPC-DAC-1995-11, Departament d'Arquitectura de Computadors, Universitat Polytecnica de Catalunya, 1995.

[SC96] F. Sánchez and J. Cortadella. Maximum-throughput software pipelining. In *Proceeding of the 2nd International Conference on Massively Parallel Computing Systems*, pages 483–490, Ischia, Italy, May 1996.

[SCF+03] B. Slechta, D. Crowe, B. Fahs, M. Fertig, G. Muthler, J. Quek, F. Spadini, S. J. Patel, and S. S. Lumetta. Dynamic optimization of micro-operations. In *Proceedings of the Ninth International Symposium on High-Performance Computer Architecture*, pages 165–178, February 2003.

[Sch87] P. B. Schneck. *Supercomputer Architecture*. Kluwer Academic Publishers, Norwell, MA, 1987.

[Sch94] J. Schutz. A 3.3V 0.6μm BiCMOS superscaler microprocessor. In *IEEE International Solid-State Circuits Conference*, pages 202–203, February 1994.

[Sch00] P. J. Schielke. *Stochastic Instruction Scheduling*. PhD thesis, Dept. of Computer Science, Rice University, May 2000.

[SD85] B. Su and S. Ding. Some experiments in global microcode compaction. In *Proceedings of the 18th Annual Workshop on Microprogramming*, pages 175–180, Pacific Grove, CA, 1985.

[SDX84] B. Su, S. Ding, and J. Xia. An improvement of trace scheduling for global microcode compaction. In *Proceedings of the 17th Workshop on Microprogramming*, New Orleans, LA, October 1984.

[SDX86] B. Su, S. Ding, and J. Xia. URPR - an extension of URCR for software pipelining. In *Proceedings of the 19th Workshop on Microprogramming*, New York, NY, October 1986.

[SDX87] B. Su, S. Ding, and J. Xia. GURPR - a method for global software pipelining. In *Proceedings of the 20th Workshop on Microprogramming*, Dept. of Computer Science, University of Colorado at Boulder, December 1987.

[SG97] J. Sánchez and A. González. Cache sensitive modulo scheduling. In *Proceedings of the 30th International Symposium of Microarchitecture*, pages 338–348, Research Triangle Park, NC, 1997.

[SG00a] J. Sánchez and A. González. The effectiveness of loop unrolling for modulo scheduling in clustered VLIW architectures. In *Proceedings of the 2000 International Conference on Parallel Processing*, page 555, 2000.

[SG00b] J. Sánchez and A. González. Instruction scheduling for clustered VLIW architectures. In *Proceedings of the 13th International Symposium on System synthesis*, pages 41–46, Madrid, Spain, 2000.

[SG00c] J. Sánchez and A. González. Modulo scheduling for a fully-distributed clustered VLIW architecture. In *Proceedings of the 33rd Annual ACM/IEEE International Symposium on Microarchitecture*, pages 124–133, Monterey, CA, 2000.

[SG01] J. Sánchez and A. González. Clustered modulo scheduling in a VLIW architecture with distributed cache. *Journal of Instruction Level Parallelism*, 3, 2001.

[SGE91] U. Schwiegelshohn, F. Gasperoni, and K. Ebcioğlu. On optimal parallelization of arbitrary loops. *Journal of Parallel and Distributed Computing*, 11(2):130–134, 1991.

[Sha72] L. E. Shar. Design and scheduling of statically configured pipelines. Technical Report 42, Digital Systems Laboratory, Stanford University, September 1972.

[SJH89] M. D. Smith, M. Johnson, and M. A. Horowitz. Limits on multiple instruction issue. In *Proceedings of the Third International Conference on Architectural Support for Programming Languages and Operating Systems (ASPLOS-III)*, pages 290–302, Boston, MA, 1989.

[SM88] M. Schlansker and M. McNamara. The Cydra 5 computer system architecture. In *Proceedings of the Sixth International Conference on Computer Design*, pages 302–306, October 1988.

[Smi88] R. E. Smith. A historical overview of computer architecture. *IEEE Annals of the History of Computing*, 10(4):277–303, 1988.

[Smi89] J. E. Smith. Dynamic instruction scheduling and the astronautics ZS-1. *IEEE Computer*, 22(7):21–35, July 1989.

[SP89] J. J. Shieh and C. Papachristou. On reordering instruction streams for pipelined computers. In *Proceedings of the 22nd Annual Workshop on Microprogramming and microarchitecture*, pages 199–206, Dublin, Ireland, 1989.

[SR00] M. Schlansker and B. R. Rau. EPIC: An Architecture for Instruction-Level Parallel Processors. Technical Report 99–111, Hewlett Packard Laboratories, February 2000.

[ST09] R. Simar and R. Tatge. How TI adopted VLIW in digital signal processors. *Solid-State Circuits Magazine, IEEE*, 1(3):10–14, Summer 2009.

[Sto93] H. S. Stone. *High-performance computer architecture.* Addison-Wesley, Reading, MA, third edition, 1993.

[SW91] B. Su and J. Wang. GURPR* - a new global software pipelining algorithm. In *Proceedings of the 24th International Symposium of Microarchitecture MICRO-24*, November 1991.

[SWX88] B. Su, J. Wang, and J. Xia. Global microcode compaction under timing constraints. In *Proceedings of the 21st Annual Workshop on Microprogramming and microarchitecture*, pages 116–118, San Diego, CA, 1988.

[Tar95] G. Tarry. Le problème de labyrinthes. *Nouvelles Annals de Math.*, 14:187, 1895.

[Tar72] R. E. Tarjan. Depth first search and linear graph algorithms. *JOURNAL of Computing*, 1(2):146–160, 1972.

[TD74] A. T. Thomas and E. S. Davidson. Scheduling for multiconfigurable pipelines. In *Proceedings of 12th Annual Allerton Conference on Circuits and System Theory*, pages 658–669, Allerton, IL, 1974.

[TF70] G.S. Tjaden and M.J. Flynn. Detection and parallel execution of independent instructions. *IEEE Transactions on Computers*, C-19(10):889–895, 1970.

[THH80] T. Temam, S. Hasegawa, and S. Hanaki. Dataflow procesor for image processing. In *Proceedings of the International Symposium on Mini and Microcomputers*, pages 52–56, 1980.

[Tho70] J. E. Thornton. *Design of a Computer: The Control Data 6600.* Glenview, IL: Scott, Foresman, 1970.

[Tie70] J. C. Tiernan. An efficient search algorithm to find the elementary circuits of a graph. *Communications of the ACM*, 13(12):722–726, 1970.

[TLS90] P. Tirumalai, M. Lee, and M. Schlansker. Parallelization of loops with exits on pipelined architectures. In *Proceedings of the 1990 Conference on Supercomputing*, pages 200–212, 1990.

[TMS] Tms320c6701 floating-point digital signal processor. http://www.ti.com/lit/ds/symlink/tms320c6701.pdf.

[TO96] M. Tremblay and J. M. O'Connor. UltraSparc I: A four-issue processor supporting multimedia. *IEEE Micro*, 16(2):42–50, April 1996.

[Tom67] R. M. Tomasulo. An efficient algorithm for exploiting multiple arithmetic units. *IBM Journal of Research and Development*, 11:25–33, 1967.

[Tou84] R. F. Touzeau. A Fortran compiler for the FPS-164 scientific computer. In *Proceedings of the 1984 SIGPLAN Symposium on Compiler construction*, pages 48–57, 1984.

[Tow76] R. A. Towle. *Control and Data Dependence for Program Transformation*. PhD thesis, Department of Computer Science, University of Illinois at Urbana-Champaign, March 1976.

[TTT81] M. Tokoro, E. Tamura, and T. Takizuka. Optimization of microprograms. *IEEE Transactions on Computers*, 30(7):491–504, 1981.

[TTTT77] M. Tokoro, E. Tamura, K. Takase, and K. Tamaru. An approach to microprogram optimization considering resource occupancy and instruction formats. In *Proceedings of the 10th Workshop on Microprogramming*, pages 92–108, October 1977.

[TTTY78] M. Tokoro, T. Takizuka, E. Tamura, and I. Yamaura. A technique of global optimization of microprograms. In *Proceedings of the 11th Annual Workshop on Microprogramming*, pages 41–50, Pacific Grove, CA, 1978.

[TW91] A. Trew and G. Wilson. *Past, present, parallel: a survey of available parallel computer systems*. Springer-Verlag, New York, NY, 1991.

[UBF+84] D. Ungar, R. Blau, P. Foley, D. Samples, and D. Patterson. Architecture of SOAR: Smalltalk on a RISC. In *Proceedings of the 11th Annual International Symposium on Computer architecture*, pages 188–197, 1984.

[Uht88] A. Uht. Requirements for optimal execution of loops with tests. In *Proceedings of the 2nd International Conference on Supercomputing*, pages 230–237, 1988.

[vdWVD+05] Jan-Willem van de Waerdt, Stamatis Vassiliadis, Sanjeev Das, Sebastian Mirolo, Chris Yen, Bill Zhong, Carlos Basto, Jean-Paul van Itegem, Dinesh Amirtharaj, Kulbhushan Kalra, Pedro Rodriguez, and Hans van Antwerpen. The TM3270 Media-Processor. In *Proceedings of the 38th Annual IEEE/ACM International Symposium on Microarchitecture*, pages 331–342, Barcelona, Spain, 2005.

[Vee86] Arthur H. Veen. Dataflow machine architecture. *ACM Computing Surveys*, 18(4):365–396, 1986.

[VK94] B. Ramakrishna Rau Vinod Kathail, Michael S. Schlansker. HPL Playdoh architecture specification: Version 1.0. Technical Report HPL-93-80, Hewlett Packard Laboratories, February 1994.

[Wal91] D. W. Wall. Limits of instruction-level parallelism. In *Proceedings of the Fourth International Conference on Architectural Support for Programming Languages and Operating Systems (ASPLOS-IV)*, pages 176–188, Santa Clara, CA, July 1991.

[Wat72] W. J. Watson. The TI ASC - a highly modular and flexible super computer architecture. In *Proceedings of the AFIPS Fall Joint Conference*, pages 34–52, 1972.

[WBHS92] N. J. Warter, J. W. Bockhaus, G. E. Haab, and K. Subramaniam. Enhanced modulo scheduling for loops with conditional branches. In *Proceedings of the 25th International Symposium of Microarchitecture*, Portland, OR, December 1992.

[WD94] S. W. White and S. Dhawan. Power2: next generation of the RISC system/6000 family. 38(5):493–502, September 1994.

[WE93] J. Wang and C. Eisenbeis. Decomposed software pipelining. Technical Report RR-1838, INRIA-Rocquencourt, France, January 1993.

[WG82] I. Watson and J. Gurd. A practical data flow computer. *IEEE Computer*, 15(2):51–57, February 1982.

[WG00] D. L. Weaver and T. Germond. *The SPARC Architecture Manual*, April 2000.

[Wil51] M. V. Wilkes. The best way to design an automatic calcualting machine. In *Proceedings of Manchester University Computer Inaugural Conference*, pages 16–18, Manchester, England, July 1951.

[Wol78] M. J. Wolfe. Techniques for improving the inherent parallelism in programs. Master's thesis, Department of Computer Science, University of Illinois at Urbana-Champaign, July 1978.

[Wol82] M. J. Wolfe. *Optimizing Supercompilers for Supercomputers*. PhD thesis, Department of Computer Science, University of Illinois at Urbana-Champaign, October 1982.

[Wol86] M. J. Wolfe. Loop skewing: The wavefront method revisited. *International Journal of Parallel Programming*, 15(4):279–293, August 1986.

[Woo78] G. Wood. On the packing of micro-operations into micro-instruction words. In *Proceedings of the 11th Annual Workshop on Microprogramming*, pages 51–55, Pacific Grove, CA, 1978.

[Woo79] G. Wood. Global optimization of microprograms through modular control constructs. In *Proceedings of the 12th Workshop on Microprogramming*, pages 1–6, 1979.

[WP95] N. J. Warter and N. Partamian. Modulo scheduling with multiple mutiple initiation intervals. In *Proceedings of the 28th International Symposium of Microarchitecture*, Ann Arbor, MI, November 1995.

[WS53] M. V. Wilkes and J. B. Stringer. Microprogramming and the design of the control cicuits in an electronic digital computer. In *Proceedings of the Cambridge Philosophical Society, Part2*, pages 230–238, April 1953.

[Yea96] K. C. Yeager. The MIPS R10000 superscalar microprocessor. *IEEE Micro*, 16(2):28–40, April 1996.

[YKM01] H. S. Yun, J. Kim, and S. M. Moon. A first step towards time optimal software pipelining of loops with control flows. *Proceedings of the 10th Conference on Compiler Construction*, 2027:182–199, 2001.

[YYF85] W. C. Yen, D. W. L. Yen, and K. S. Fu. Data coherence problem in a multicache system. *IEEE Transactions on Computers*, C-34(1):56–65, January 1985.

Index

A. Aiken et al., *Instruction Level Parallelism*,
DOI 10.1007/978-1-4899-7797-7

L

M

V

W

X

Z

Printed in the United States
By Bookmasters